Theophrastus, James G. Wood, George J. Symons

Theophrastus of Eresus on Winds and on Weather Signs

Theophrastus, James G. Wood, George J. Symons

Theophrastus of Eresus on Winds and on Weather Signs

ISBN/EAN: 9783337343507

Printed in Europe, USA, Canada, Australia, Japan

Cover: Foto ©Andreas Hilbeck / pixelio.de

More available books at **www.hansebooks.com**

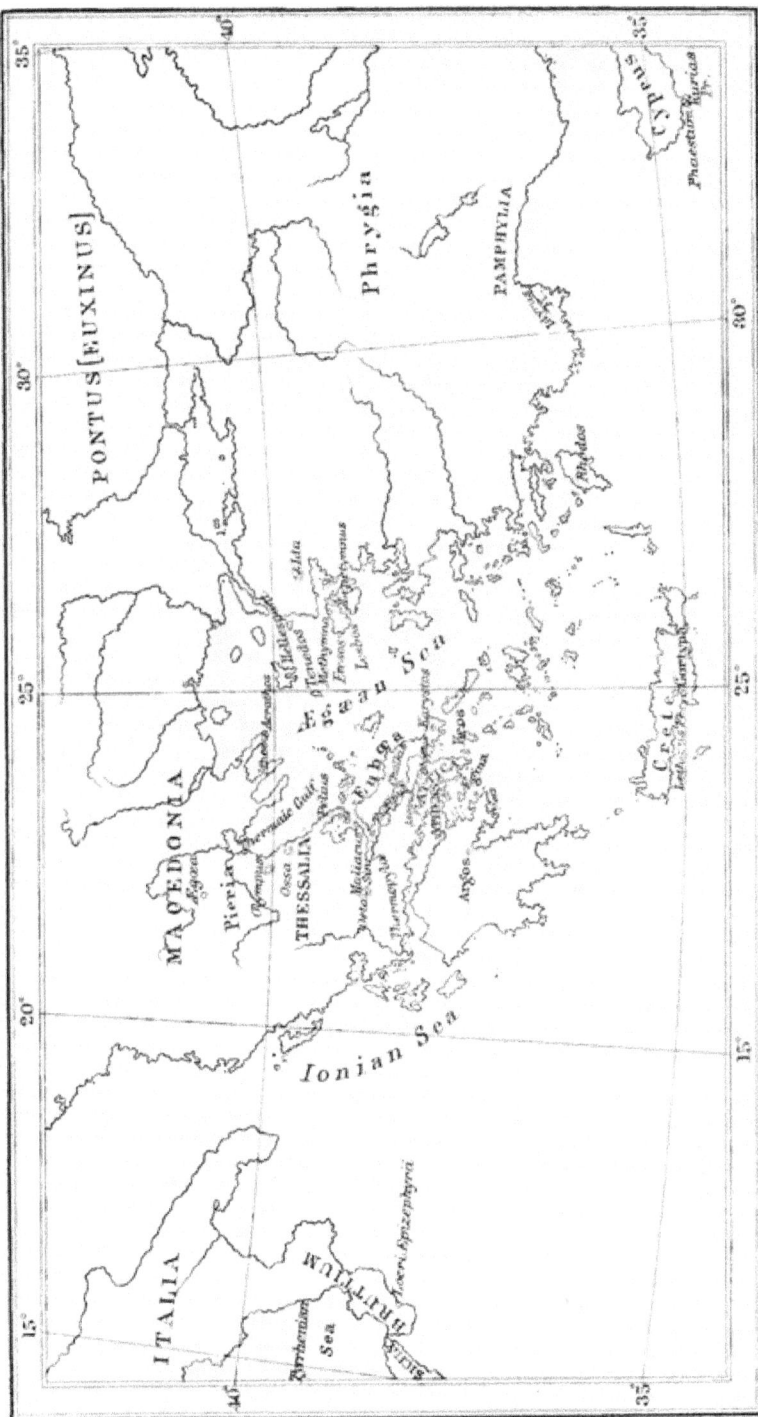

SOUTHERN EUROPE – SHOWING PLACES NAMED BY THEOPHRASTUS.

VICINITY OF ATHENS—SHOWING PLACES NAMED BY THEOPHRASTUS.

MAP B.

THEOPHRASTUS

OF ERESUS

ON

WINDS

AND ON

WEATHER SIGNS.

TRANSLATED, WITH AN INTRODUCTION AND NOTES,

AND AN APPENDIX

ON THE DIRECTION, NUMBER AND NOMENCLATURE OF THE WINDS
IN CLASSICAL AND LATER TIMES,

BY

JAS. G. WOOD, M.A., LL.B., F.G.S.,

*Of Lincoln's Inn, Barrister-at-Law, and sometime Fellow
of Emmanuel College, Cambridge,*

AND EDITED BY

G. J. SYMONS, F.R.S.,

*Chevalier de la Légion d'Honneur,
Secretary of the Royal Meteorological Society,
&c., &c.*

LONDON:

EDWARD STANFORD,

26 & 27 COCKSPUR STREET, CHARING CROSS.

1894.

CONTENTS.

ILLUSTRATIONS.

LIST OF SUBSCRIBERS.

The Royal Society.
The Royal Observatory, Greenwich.
The Royal Observatory, Edinburgh.
The Royal Meteorological Society.
The Meteorological Council.
Weather Bureau, U. S. Department of Agriculture.
Observatoire Physique Central, St. Petersburg.
Königliches Preussisches Meteorologisches Inst., Berlin.
K. K. Central-Anstalt für Meteorologie, Vienna.

Appach, Miss.
Bayard, F. C.
Bell, Major.
Bicknell, P.
Blackmore, R. D.
Boys, Rev. H. A.
Boys, W.
Brook, C. L.
Brown, Rev. D. Dixon.
Buchanan, G.
Case, C. A.
Chandler, A.
De Rance, C. E.
Dixon, G.
Dudgeon, P.
Eaton, E. M.
Evans, F. G.
Filgate, T. F.
Ford, A. L.
Hann, Dr. J.
Harrington, Prof.
Hatfield, W.
Hellmann, Prof. Dr.
Hopkinson, J.
Howard, W. Dillworth (2 copies)
Inwards, R.
Jebb, J. R.
Jenkyns, Lady.
Latham, Baldwin.
Lippincott, R. C. C. (2 copies).
Macdonald, Rev. J. A.
Mace, J. E.
Maclear, Admiral.
Marten, E. B.

Mawley, E.
Mitchell, Rev. J. C.
Morton, Dr. J.
Mossman, R. C. (2 copies).
Parbury, A. F.
Plenderleath, Rev. W. C.
Prince, C. L.
Pringle, C. S.
Rotch, A. L.
Russell, H. C.
Scriven, R. G.
Sidebottom, J.
Silver, S. W.
Simpson, E.
Smith, B. Woodd.
Snowden, Rev. H. C. V.
Southall, H.
Stubs, P.
Sturt, Col. N. G.
Topley, W.
Trotter, J.
Vaughan, Cedric.
Wallis, H. Sowerby.
Ward, Col.
Ward, R. De C.
Watkins, J.
Welby-Gregory, Sir. W., Bart.
Wesley & Son, Messrs.
Williams, Dr. Theodore.
Williamson, B.
Wohlleben, T.
Wood, J. G. (6 copies).
Yool, H.

PREFACE.

It is, perhaps, desirable to say a few words respecting the circumstances which have led to the preparation of this book. For nearly thirty years I have been a persistent searcher for old works upon Meteorology, partly like other hunters, from the universal desire to capture something : but I hope more with two other objects, (1) to form a centre where much of the meteorological literature of past ages could be consulted, and, as far as my own small efforts could go, to insure its permanent preservation : (2) to learn from it all that I could as to the growth of our knowledge of meteorology.

I have always had before me, but I fear only as a vision never to be realised, the translation and publication of, at any rate, the most important of these works. Aristotle's Meteorology has been translated into English (but copies are very scarce), and there is the excellent French translation of Barthélemy St. Hilaire, and it is a large work. I thought, therefore, that it would be more prudent to begin with the smaller works of Aristotle's favourite pupil Theophrastus, which existed only in Greek and Latin ; and in the *Meteorological Magazine* for July, 1892, I asked whether any one would volunteer to prepare the translation, promising that in that event I would undertake the cost of publication. I was favoured with no fewer than four offers, and accepted the first. Of *how* Mr. Wood has carried out his "congenial labour" it would be presumptuous for me to say one word, but it cannot be wrong to express my hearty thanks for his help, also for that proffered by others which it was not necessary to accept.

"The Cobham Journals," which Miss E. A. Ormerod kindly published, "Cowe's Meteorological Journal," and "Merle's MS., 1337-1344" have been earlier efforts in a somewhat similar direction; but in Theophrastus we go back more than 2,000 years and to a totally different class of writing.

I had thought of preparing a brief commentary upon the two papers, but found that it could not be "brief"; and it is so uncertain how much was original on the part of Theophrastus, how much he had derived from Aristotle, how much Aristotle owed to Herodotus, and Herodotus to his predecessors, that it seemed better not to attempt it.

I hope that the two maps and the index giving the position of every place mentioned will facilitate the understanding of some of Theophrastus' arguments.

Although I see that Mr. Wood has, in the Appendix, expressed his indebtedness to the Rev. Padre Denza for the care and success with which he supplied both photographs and casts from the "Table of the Winds" at the Vatican, I trust that there is no impropriety in my adding my own thanks not only for that help, but for the promptitude which Padre Denza ever shows to assist those who try to advance that science for which he has done so much.

G. J. S.

62, CAMDEN SQUARE, N.W.
June 2nd, 1894.

INTRODUCTION.

THE author of the following papers, generally known to us as Theophrastus, was, in his early days, called Tyrtamus: and the name Theophrastus (the divine speaker) is said to have been given to him by Aristotle on account of his eloquence.[*]

He was born at Eresus in Lesbos in or before B.C. 371; and died B.C. 287. Coming early to Athens he became a pupil of Plato; but after the latter's death (B.C. 347) became attached to Aristotle, probably during the time when the young Alexander of Macedon also was his pupil. The closeness of the intimacy which was then begun, and the regard of the master for the pupil, are evidenced by the fact that Aristotle on his death (B.C. 323) appointed Theophrastus one of his executors, and bequeathed to him his literary property and his Library, including Aristotle's own works, which eventually passed from the hands of Theophrastus to Neleus of Scepsis, and from him to Ptolemy Philadelphus, who deposited them at Alexandria.

At Aristotle's death Theophrastus succeeded also to the direction of the Lyceum, which the former had founded at Athens as the seat of his peripatetic system of scholastic and philosophic disputation. The Lyceum was continued by Theophrastus during his life, and bequeathed by him to Straton and others on condition that it should remain and be carried on as a philosophical college. Here, until later developments and sub-divisions arose, the Aristotelian philosophy maintained its home; as did the Platonic, side by side with it, at the Academy; and here Theophrastus was the centre of a numerous body of pupils in the various branches of study which the unfailing energy of its founder had made his own. Among

---

[*] Diog. Laert., v., 38.

these were such men as the poet Menander and the orator Deinarchus.*

Whether the labours at the Lyceum interfered with literary work, or whatever else was the reason, the works of Theophrastus which have come down to us are not numerous; and the majority of them are known to us only by fragments. Many of the others appear rather to be outlines preparatory to complete works than complete works in themselves; or material to be amplified and elaborated in lectures to, or disputations with, his pupils; corresponding somewhat to the division of Aristotle's works which Dr. Donaldson speaks of as the " hypomnematic works, or draughts and notes of books which " Cicero calls 'commentarii' as distinguished from the syntag- " matic or complete and formal treatises."

To the class of less formal works the two short treatises (or, as we should term them, papers) now for the first time presented in an English form appear to belong.

Of these two papers, the one " On Winds " may be said to be the more nearly complete; but it is far from being exhaustive. At its outset† and throughout, it postulates or assumes an acquaintance with the Aristotelian theory of " the winds," and of the origin and nature of " wind," either as developed in some earlier treatise of our author, or in the works of his great master. It deals with the effects of wind, and the explanation of concurrent and consequent phenomena, rather than the cause of wind itself or the origin of particular winds. The earlier parts follow fairly regular divisions of the subject; but towards the end there are evidences of haste and incompleteness, such as we find still more apparent in the next paper.

---

* The authorities for the particulars of Theophrastus' life will be found collected in Muller and Donaldson's Literature of Greece, Vols. II. and III., titt., " Aristotle " and " Schools of Philosophy," from which the above short notice of our author is practically derived.

† The reference in the very opening lines, where it is said that the origin of the physical constitution of the winds " has been already considered " (τεθεώρηται πρότερον), is probably to the passages in Arist Met.; lib. i., cap. 13, and lib. ii., cap. 4, which I have dealt with later on; and, if it is genuine, to de Mundo, cap. iv. But, unless indeed the latter work is our author's, and not Aristotle's, we possess nothing of Theophrastus to which that passage refers.

The paper "On Weather Signs" has even less pretension to be considered a complete or original treatise on its subject. It reads as if it were a collection of notes taken of a lecture delivered, or notes to form the basis of, and to be expanded in, a lecture to be delivered. There are so many passages which can be traced to the *Meteorologica*, or the *Problemata*, of Aristotle, that it may well be that it had its origin at the time when Theophrastus was the pupil of Aristotle, and be a *resumé* of what he heard from his master. This theory receives some support from the reference to the "Drawing" or diagram mentioned in section 35 which, as Theophrastus has given us no diagram of his own, can hardly be other than that prepared by Aristotle and more fully described hereafter. That this was so is confirmed by a short paper which Aristotle has left us, "On the position and names of the Winds," in which, after setting down their names, he proceeds thus:—"And I have written down for thee at the foot their position, how they are disposed, and from what quarters they blow; and have made a diagram of the circle of the earth, so that they may be set forth before thine eyes." It can scarcely be doubted that this is again the same, or a copy of the same, diagram, and that the person addressed was Theophrastus.*

As an instance of the incompleteness of this paper, it will be noticed that, although in sections 1 and 2 our author has with some pains laid down certain propositions as to the diurnal risings and settings of the stars, and referred to certain signs as occurring at those times, he nowhere explains what those signs are. In fact, he has told us nothing to which those propositions lead; the work would be equally complete without them. In sections 7 and 8 the setting of the Pleiades is referred to; but that is the annual, not the diurnal, setting. So also the risings of Sirius and Arcturus referred to in section 23 are the annual risings. Again in section 57 there is a general reference to "indications of the times of the appearance of stars;" but this is the annual, not the diurnal, reappearance. This latter passage indeed seems to indicate that a further portion of the work, in

---

* See Appendix, p. 79.

which such indications were, or were intended to be, dealt with,
has either been lost, or was never written.

To appreciate fully the theories upon which these writings
of Theophrastus are based, or which these writings themselves
put forward, is difficult at the present day; and difficult for this
reason, that it is almost impossible to place ourselves on the
same standpoint, as men to whom the rotation of the earth
about the sun was as yet a dream, or a theory to be disclosed
only to a select few, and to whom the nature of heat, and the
composition of the gaseous, fluid, and solid forms of matter
which go to make up our world, were absolutely unknown, though
matters of constant speculation.* We must for the time accept
(as a basis) the Aristotelian theory that the elemental principles
(ἀρχαί) are the four fundamental properties of "Nature," namely,
"the hot," "the cold," "the dry," and "the moist"; that of
these the mixture of hot and dry produces fire; that of hot and
moist produces air (air being as it were a vapour) ; that of cold
and dry produces the earth; and that of cold and moist produces
water.† But even this theory, stated in terms apparently so

---

* I may apply to my own case the following words of Lucretius (Bk. i.,
137—146), in which he expressed the difficulty of representing even in his
day the philosophy of Greece to the students of Rome – a difficulty which
the still greater distance of time and change of thought and expression has
proportionately increased : —
    " Nec me animi fallit Graiorum obscura reperta
    difficile inlustrare Latinis versibus esse,
    multa novis verbis praesertim quom sit agendum
    propter egestatem linguae et rerum novitatem;
    sed tua me virtus tamen, et sperata voluptas
    suavis amicitiae, quemvis sufferre laborem
    suadet, et inducit noctes vigilare serenas,
    quaerentem dictis quibus et quo carmine demum
    clara tuae possim praepandere lumina menti,
    res quibus occultas penitus convisere possis."
† See Arist. de Generat. II., 3, § 2. It may be useful for future reference
to tabulate the formulae as follows : -
Hot + dry = fire (representing physical force).
Hot + moist = air (representing the gaseous form of matter).
Cold + dry = earth (representing the solid form of matter).
Cold + moist = water (representing the fluid form of matter).
It must be understood that "moist" does not necessarily in itself involve

simple, it is difficult for us really to grasp, so foreign is it to our habits of thought founded on more accurate knowledge. But it may be that underlying it, or attempted to be thereby expressed, is the proposition that all "matter" is in itself homogeneous and of one simple substance, and that what are known to us of the present day as elementary bodies differ, as between themselves, not in essence, but as being forms of the same one elemental matter presented to us under conditions varying in each case, and due to the action of physical forces acting in combination or opposition; and that, as a consequence, each of such elementary bodies is by a due variation of the conditions resoluble into, or interchangeable with, each and every of the others.

Indeed, this proposition is almost stated in terms in Arist., Met., i., 3, when he says that "fire, air, water, and earth are all "derivable from each other; and each of them pre-exists "potentially in each of the others."

Adopting then for this purpose this theory of the elemental properties of nature, we must now proceed to state briefly the Aristotelian theory of the origin of wind, and what "a wind" was understood to be.

This we learn principally from Aristotle's *Meteorologica*, lib. i., cap. 4 & 13, and lib. ii., cap. 4. To have translated these passages at length would have added too much to the bulk of this work. I have therefore set out only the more important propositions, paraphrasing the greater part to render them more intelligible, but giving (as quotations) such parts as I thought necessary to treat more literally.

The initial principle (ἀρχή) of wind is "fire;" but by "fire" we are to understand not mere combustion, but a force capable of modifying the conditions of matter, and imposing on the modified particles of matter the tendency to ascend from the earth's surface towards that outermost region or envelope of our system, which is the habitat of fire. The manner of modi-

---

the idea of dampness due to the presence of water. It rather represents "fluidity," being the property common to air and to water; but still a fluidity the opposite of "dryness" and not the opposite of "solidity."

fication is ἀναθυμίασις, which, for want of a better English word, I have translated "sublimation." When the subject of modification is matter in a fluid state (or the "moist") the modified form is "vapour" (ἀτμίς); and in analogy to this, when the subject of modification is in the solid state the modified form is "smoke" (καπνός), and, for want again of a better word (as well in Greek as in English), this term is applied to the result of all sublimation other than vapour. And so there are two kinds of "sublimation;" one the moist or "vaporiform" (ἀτμιδώδης), the other the dry or "fumiform" (καπνώδης). But "the moist" does not exist without "the dry," nor "the dry" without "the moist;" but the excess of the one over the other determines the quality of the form.

"As the sun moves in its orbit it draws up the moist when it "approaches; but as it recedes, the vapour which has been "drawn up is again condensed by the cold into water." This is in fact the moist sublimation which, owing to the excess of the moist over the dry, is the "principle" of "water in the "form of rain." The "dry sublimation" is the principle and physical origin of all the winds.

Some say that what is called the air, when it is in a state of motion and flux, is wind; and this same matter when solidified is cloud and rain; and that physically rain and wind are the same: and wind is a movement of air; and all the winds are but one wind; and they do not differ from each other, although by reason of the difference of the places from and to which they blow they seem to be different. But this is all wrong; for one might just as well say that all rivers are but one river. But our notion of a river is not merely a quantity of water, however great, flowing just anyhow; but it must have a spring in a particular place for its source, and it must have a defined channel. So each wind, whether the N. or the S. or any other wind, has its particular place of origin and its defined direction; and they are not merely disturbances of the whole air, or convertible into each other.

Now, the air is derived from vapour and smoke.* Vapour is

* See also Arist. de Generatione II., 3, § 2.

moist and cold; smoke is hot and dry; and these two contribute (καθάπερ ἐκ συμβολῶν, not συμβόλων) to the formation of air which is moist and hot, or, stating it as a formula,

"Vapour" = moist + cold;

"Smoke " = hot + dry;

"Air " = moist + hot.

In the formation of air (and so of wind) there must therefore be an excess of moist in the "vapour," and an excess of hot in the "smoke." Sublimation goes on more or less continuously, varying in amount; and the vaporiform sublimation is sometimes greatly in excess, and at other times the dry and fumiform: and hence arise the variations, in different seasons and between different places, between droughts and rains, calms and winds; the forms of sublimation being interchangeable and variable both as to time and as to place.

After showers, a wind generally occurs where the showers have been; and winds drop when rain has come; and it must needs be so, on the principles already stated. For when it has rained, the earth, beginning to dry up under the influence of the heat within and the heat above, is in a state of sublimation; and this (i.e., the combination of hot and dry) we have seen to be the "body " (σῶμα) or "material " (ὕλη) of wind.

A wind then is an excess of the dry sublimation from the earth set in motion around the earth; and the origin of its movement is from above: the origin of its matter and its production is from below.*

From these passages we learn the following propositions:— first, that the winds are separate and distinct entities flowing in definite courses, and not mere movements of the same air hither and thither; secondly, that to produce wind, matter has to be formed, and the more matter the greater the wind; and this " matter " is derived from the earth, and is distinct from

---

* This is sufficiently accurate for the present purpose; but to render it intelligible if rendered in the strict language of Aristotelian philosophy, it would be necessary to go into the subtleties of the movent principle and the generative principle in a manner far outside the scope of the present work.

vapour; and just as in the case of "moist sublimation," when the involved heat is discharged in the higher regions of our system, the remaining matter becomes solidified as rain; so in the case of dry sublimation, when the involved heat is similarly discharged, the remaining matter becomes solidified as wind; and the motion of wind is due to the over-production of the matter, and the necessarily consequent effort to restore equilibrium.

The difficulty of representing these ideas in our language is increased by the fact (familiar to my scholarly readers) that our one word "Wind" is a very insufficient equivalent for the three Greek words ἄνεμος, πνεῦμα and πνοή. The distinction (carefully observed in the works before us) between the ideas conveyed by those three words may be to some extent marked by speaking of ἄνεμος as "a Wind;" or (with the definite article) in the plural as "the Winds;" as being definite concrete entities; while πνεῦμα is represented by "Wind" in the indefinite or abstract; and πνοή by current or wind movement. But the distinction may appear clearer if, regarding (as Aristotle and Theophrastus undoubtedly did) ἄνεμος (a Wind) as in form analogous to a river, we understand πνεῦμα ("Wind") to be the stream flowing in that river; and πνοή as the current or movement of the stream.

I have in the Appendix attempted to give a review of the changes which are to be noticed from the time of the early Greek poets down to the middle ages, both in the number of recognised distinct winds and in their names, and in the relative positions to which they were assigned.

For the present purpose it is sufficient to say that in order to render the translation more readable, I have throughout (except in a few instances when the reason for the exception is obvious) avoided the use of the Greek names of the winds, and in place of such names have spoken of the N. wind, S.E. wind, and so on; but it will be understood from what I have said on the subject in the Appendix that, except in the case of the winds from the four cardinal points (N. S. E. W.), the winds named by Theophrastus are only more or less approximately to be referred to the particular compass points to which I have referred them, according to the following table:—

COMPASS.	GREEK.
N.	Boreas.
N.N.E.	Meses.
E.N.E.	Kaikias.
E.	Apeliotes.
E.S.E.	Euros.
S.S.E.	Phœnikias or Euronotos.
S.	Notos.
S.S.W.	Libonotos.
W.S.W.	Lips.
W.	Zephyros.
W.N.W.	Argestes.
N.N.W.	Thraskias.
N. by W.	Aparctias.

The principal editions of the works of Theophrastus are those of Schneider (Leipsig, 1818) and Wimmer (Paris, 1866). The following translation has been prepared with the help of both those editions. Upon the whole I have followed Schneider's the more closely; but have not hesitated to select in particular cases whichever reading appeared to me to give the better sense. The few instances in which I have departed from both will be found duly noticed.

As both editions afford full information for those who wish to enter on a critical study of the text, I have thought it unnecessary, if not foreign to the object of the present work, to enter into such matter.

Both editions contain also a Latin translation apparently drawn from a common source; but in cases of real difficulty I have found it of little service. In such cases the translator seems to have contented himself with replacing each Greek

word by a Latin one, instead of ascertaining the meaning of
the Greek and clothing it in a Latin dress.

For convenience of reference I have retained the numbering
of the sections as given in those editions, though in some places
the divisions of the sections do not correspond with the actual
divisions of the subject matter. This will account for the
numerals not being always at the beginning of a section.

In parting with the congenial labour of many hours taken
from the intervals of professional work, and handing it over to
the kindly criticism of English readers, I am conscious that much
more might have been done on my part towards presenting
these specimens of early meteorological work in a more attrac-
tive and perhaps more useful form. But having presumed to
undertake the task kindly entrusted to my hands, I have no
right to ask for, though I hope I may receive, consideration for
such imperfections as may be found in its performance; and I
desire to say nothing more of my own share of the work.

I feel, however, that there may be some, in the latter end of
this nineteenth century, to whom it may seem futile and out-of-
date to have thus unearthed the speculations of an Athenian
philosopher of a bye-gone age; and who, satisfied with a
cursory perusal of the following pages, may dismiss them with
contempt, as unworthy of consideration by disciples of the
modern school. To these I would say this:—The brightness of
the noonday sun has never shone upon us, but it has first been
preceded by the grey light of the growing dawn, little as many
of us may know of, or care for, the beauty of those earlier hours.
No great river has ever rolled itself on into the mighty sea, but it
has first, scarcely seen or thought of, groped its way in the shade
and obscurity of the mountain beds that enclose its tiny rills,
and then by degrees " slowly broadened down," until at last its
swelling waters teem with the busy works of men, who, accustomed day by day to look on it in its fulness, and accept its aid
as a matter of course, little think of the far-off and small but
necessary beginnings which have conduced to such an end. So,
too, in the case of science, there must be beginnings. How far
from the beginning Theophrastus was—how near to the end we

are—who shall say? Let us never forget the debt we owe to those who first set flowing the streams of knowledge which have united and widened out into the fuller possessions we enjoy ; or suppose that without the early labours of such pioneers into the dark recesses of the mysteries of nature, we could ever have walked firmly along the broad highway that seems to us so plain and smooth.

It may, indeed, be that the more we study such works as those before us, the more we shall find that there are, in them, treasures of thought, of observation, and of expression which may yet enrich us, if we will but use them ; or may again read, in the mistakes of the past, a warning that, as then, so in the present, the finest and most highly trained of human intellects are capable of error : or at least we shall learn something of a sympathy and a fellowship, unbroken throughout the ages of time, arising from the common desire that links the labours of the Lyceum to the labours of to-day—the desire to Know—a desire which, manifesting itself in the earliest moments of the history of our race, has ever remained, and will remain, unsatisfied and insatiable until at last (if I may borrow the language of a master of modern literature) "Our tiny cockboat of knowledge "is swallowed up in the mighty ocean of GOD's Truth."

J. G. W.

115, SUTHERLAND AVENUE, W.
May, 1894.

ON WINDS.

FROM what elements, in what manner, and through what causes the physical constitution of the winds derives its origin has been already considered.[1] We must now endeavour to show that each wind is accompanied by forces and other conditions in due and fixed relation to itself; and that such conditions in fact differentiate the winds one from another.

Now, the differences that exist involve, and consist of, such conditions as the following:—for instance, greater or less volume, cold, heat, or (in more general terms) storm or calm, wet weather or clear weather; and, again, their frequency or infrequency; their occurrence season by season, or not at all seasons; and whether they are continuous and uniform, or intermittent and variable. In a word, they involve every condition that arises in the heavens, the air, the earth, and the sea by reason of the blowing of the wind. Our enquiries in fact follow the same lines and concern the same matters as do the studies of animals and plants.

Now, as each wind has its own particular place of origin, and this is, as it were, of its essence, it is from this that the distinctive features and peculiar forces of each generally arise; such as, in the first place, the greater or less volume; higher or lower temperature; greater or less amount, and [secondly] the majority of other physical conditions.

Winds from opposite quarters have, at the same time, both

---

[1] See Introduction, p. 10.

identical and opposite characteristics; and there is no inconsistency in this. Take, for instance, the N. wind and the S. wind. They both are strong winds, and blow longer than any others; and this is because the greatest amount of air is compressed towards the North, and towards the South; those parts being to the right and the left of the path of the sun, from its rising to its setting; for here² the air is expelled by the power of the sun; and so the air [there] has the greatest density, and the greatest amount of cloud. A great quantity of air being thus collected at each point, a greater and more continuous flow of air thence occurs with greater frequency; and from these causes, these winds derive volume, continuance and duration, and other such conditions.

But the coldness of the one, and the heat of the other, appear to be most clearly due to the place of origin of the particular wind; for the Northern parts of the world are cold: the Southern parts warm: and the air that flows from either quarter has the corresponding character. Now, the less open the surrounding district is, so in proportion is the current less diffuse; while that which is borne through a narrow space and with more violence is colder; but that which is poured abroad into the wider space beyond becomes more moderate in its rate of motion.³ For which reason also the S. wind is colder there than it is with us; and even, as some say, colder than the N. wind.⁴

---

² That is in the path of the sun.

And therefore less cold. See this idea further elaborated in §§ 19, 59 below. The foundation of it seems to be that a wind starts from a point, and therefore at first occupies a small space; but broadens out as it proceeds. If that were the case a continually decreasing part of what was in its origin a S. wind would continue such; while the rest would be constantly radiating off to the right and left. The difficulties with which such a theory is fraught are too obvious for further discussion.

⁴ Theophrastus here endeavours to solve the question put, but not answered, in Arist., Probl. xxvi., 16.

But the change [in its temperature] makes itself more appreciable as the place is warm to begin with.

Indeed this [variability] is common practically to all winds. *4* Whether the wind from either [of two opposite quarters] produces wet or clear weather, whether it is squally or steady, recurrent or continuous, uneven or even, or again its intensity in one case at its commencement, in another as it ceases, are matters which are more directly referable to the distance apart of the places [of origin and of observation]. For wherever any particular wind blows from, there it is accompanied by clear weather; but to whatever place it impels the air there it is accompanied by clouds and rain. This is the reason why the N. wind, and even more so the monsoon, brings rain to those who live towards the South and the sunrising; while the S. wind, and generally speaking all winds which blow from that quarter, bring rain to those who live towards the North.

And, in this connection, it is not of slight, but of the greatest, *5* importance that the places [where such effects are observed] should have a sufficient elevation. Wherever the clouds strike and take up a position, there also is a source of rainfall. For which reason, of several places close together, some are wet in the presence of some winds; others in the presence of others; rain however has been elsewhere spoken of at greater length.

It is from the same cause that the N. wind is strong immediately it begins to blow; while the S. wind is strong as it is leaving off; on which facts is founded the proverbial advice about sailing.[5] For the former immediately as it were attacks those who dwell in the North; but the latter stands far aloof:

See Arist., Probl. xxvi., 21 & 47. In the latter passage a different reason is given. The proverb may be rendered—

" 'Tis well to sail,
When the South winds begin to blow,
And when the North winds fail."

but when volume has been gathered, then comes the rush from
afar, although after a longer delay. And so conversely in
Egypt and places thereabouts the S. wind is strong at the
beginning: whence there they reverse the proverb.

6   In like manner, with them, the S. wind particularly exhibits
the characteristics of recurrence, steadiness, continuousness and
regularity, for such is always the character of each wind among
those who are near its place [of origin]; but when it reaches
those who are afar, it is irregular and disorganised.[6]

The causes which have been mentioned must be understood
to be the causes of these latter facts as well; and they are
plainly active even in other places of less extent and less distant
from each other [than Greece and Egypt]; although this might
seem unlikely. [But it is not so really]; for the S. wind is
always accompanied by clear weather in the place of its origin;
but the N. wind, whenever there is a great storm, produces
cloud in the parts near [its place of origin], but clear weather
beyond.   And the cause of this is, that by reason of its force it
sets in motion a great quantity of air; but the congelation
produced by it takes place before it can effect the propulsion of
the air so set in motion, and the clouds remain fixed by reason
of their weight; and so the force of the wind, rather than its
low temperature, is passed on to the parts beyond, and further
in advance; and produces the result we have mentioned.   The
S. wind on the other hand having substance[7] to a less degree,
and not congealing it, but propelling it, brings clear weather to
those near its place of origin; but it is always more rainy
beyond, and blows with force rather when ceasing, than when
commencing, because it propels before it only little air at the
beginning, but more as it advances;[8] and the air, by being

---

[6] See above note ([1]) on § 3. Here, as just before, the author is using a
military metaphor.

[7] See Introduction, p. 15.

[8] This seems entirely inconsistent with the argument in § 3.

gathered together, becomes cloudy, and by condensation becomes moist.

Moreover, it makes a difference whether the active principle[9] at the beginning be greater or less; for when that is small, the wind is accompanied by clear weather; and when it is great, it produces cloudy and wet weather; because [in the latter case] it compresses together a greater volume of air.

Now, they say that it is not true, but false, that the S. wind 8 [as some assert][10] does not blow at all in that part of Egypt which is near the sea, nor for the space of a day and a night's journey therefrom, while it blows fresh in the parts above Memphis, and likewise in places as far from the sea as the distance just mentioned. However, it clearly does not blow as much there, but less;[11] and the reason of this is, that lower Egypt lies low and flat; so that there the wind may pass overhead, while Upper Egypt is more elevated. Indeed the proximity of its place of origin demands that the force of the winds should be exhibited there; for, such phenomena as these, which happen according to the course of nature, are mostly referable to local causes.

And these winds continue throughout accompanied by cloudy weather, or by fine weather (as the case may be), according to what has just been said.

The proposition that the N. wind succeeds the S. wind, but 9 that the S. wind does not succeed the N. wind, must be considered with reference to the principle which assigns particular

---

[9] See Introd., p. 13.

[10] This was assumed as a fact, and the cause of the supposed fact enquired for in Arist., Probl. xxvi., 46.

[11] I adopt Schneider's and Wimmer's reading, οὐ μὴν ἴσως γε, ἀλλ ἔλαττον πνεῖ. for the common reading οὐ μὴν ἀλλά γε ἔλαττον πνεῖ. which, however, Wimmer has followed in his translation, " Verumtamen fortasse remissius tantum spirat," or, " However it may be that it only blows less."

phenomena to particular localities. For this law operates with
us, and generally with all those who live towards the North;
but with those who live towards the South the converse takes
place. The cause, however, is the same in both cases. For to
the former the N. wind is close at hand; and so is the S. wind
to the latter; so that they produce sensation directly that
they begin to blow; but make their way slowly to parts
farther on.

10   Now, the Northerly and Southerly winds being, as has been
said,[12] the most frequent, each of them is subject to a fixed
rule, as it were, determining the periods during which they
generally blow in regular order. The N. winds blow both in
Winter and Summer, and in the late Autumn until just before
the close.[13] Southerly winds blow in Winter and at the commencement of Spring, and the end of the late Autumn. For
the motion of the sun co-operates with each, and equilibrium is
restored[14] by the air flowing back again. For, whatever amount
of air may have been expelled [by N. winds] during the Winter
(and the N. winds generally blow then more frequently than
the S. winds) and again before Summer by the monsoons[15] and
succeeding winds, it is given back to these parts in the Spring
and at the close of the late Autumn, and about the setting of
11 the Pleiades in due course. And so it is that the very fact,
the assumed non-existence of which has caused some to wonder
why it is that there are monsoons from the North but not also

---

[12] Cf. § 2 above. Also in Arist., Met. ii., 4, it is said, " The N. winds and
the S. winds occur most frequently of all."

[13] The passage as to the duration of the N. wind is lost in the original;
but I have adopted Schneider's restoration of the missing portion.

[14] More literally "compensation takes place."

[15] I have for convenience adopted this word as an equivalent for ἐτήσιαι,
" the annual winds"; but it is not to be understood that these had all the
other characteristics of the true monsoon, besides that of annual recurrence.
They were N. winds.

from the South, appears really to happen. For the S. winds
of the Spring (which they call " white S. winds " from their
being usually accompanied by clear weather) are, as it were,
monsoons; but at the same time, by reason of their being
removed far off from us, they have not been recognised as such;
while the N. wind is immediately present to us.

We will now consider the nature of the Monsoon.

Why it blows at this particular season, and for a particular
number of days, and why it ceases as the day closes in, and
almost universally does not blow at night, are to be explained on
the following principles. The movement of the air is caused
by the melting of the snow. When, then, the sun begins to
break up the frost, and acquire the mastery, the "precursors"
blow; and then follows the monsoon. And the cause of its 12
ceasing with the decline of the sun, and not blowing at night,
is that the snow ceases to thaw as the sun goes down, and does
not melt at all at night when the sun has set. However, [the
monsoon] does blow sometimes [at night] when the thaw has
been greater than usual; for this must be taken to be the cause
of this exceptional occurrence. For at one time it is strong,
and continuous; at another weaker and intermittent; and this
is because the thaw is irregular. And the moment varies as
the mass of matter.[16] And it may be that this irregularity is
due to local causes, such as proximity and distance and other
variations.

If, then, it is true (as some and particularly the dwellers in 13
Crete say) that the winters are more severe, and more snow falls
than formerly—(as proof of which they allege that formerly
the hills were inhabited and produced both corn and fruit, the

---

[16] This seems to be a quotation from some mathematical work. See Introd.,
page 15, as to ὕλη, or " matter."

land having been planted and cultivated for that purpose; that
there are in fact on the hills of the Ida range and on others,
plateaus of considerable extent of which now-a-days they culti-
vate not one, because they are unproductive; while formerly
as has been said they not only cultivated them, but also dwelt
upon them so that the Island had a large population; and that
at that time showers occurred, but much snow and storm did
not)—if, I repeat, this is true which they allege, it follows that the monsoon also has greater duration [now than
formerly].[17]

14   But if the monsoon did ever fail altogether, and Aristæus (as
they tell us in the Mythologies) regained it by performing those
celebrated sacrifices to Zeus in Keos,[18] it would follow that the
parts exposed to the weather were not then so subject, as they
are now, to shower and to snow. But if rain and snow are
liable to variations, either subject or not subject to some fixed
law, there would be, synchronously with these variations,
cessation and mutation of the winds.

And it would seem strange if those in the South had not
some such relief as this, year by year; considering that their
situation is so much hotter.

This then is plain except . . . . the fruit; . . . . some are
beforehand; others are insensible. These matters must be
enquired into.[19]

15   Now if the genesis of all the winds be the same, and be

---

[17] Cf. Arist., Met. ii., 5, where it is maintained that the monsoon was caused
by the melting of the snow after the summer solstice. Hence it would follow
that the more the snow the greater the amount of liquefaction, and the
greater the consequent air-movement.

[18] The story of Aristæus and the bees, in Virg. Georg., Bk. iv., is familiar.
That of his regaining the mild climate of Keos will be found in Apollonius
Rhodius, Argon ii., 500 et seq.

[19] This passage is mutilated; and what is left is apparently corrupt. It
seems hopeless to attempt to restore it.

produced by the same agents through the acquisition of some
matter, the sun would probably be that which produces them.
But perhaps that is not absolutely true ; but rather that
sublimation[20] is the producing cause, and the sun is, as it were,
the co-operative. But the sun appears both to set the winds in
motion, and to lull them to rest, at its rising ; and so [at that
time] they often increase or die away. This however is not
universally true ; but in whatever cases it happens, the following
must be understood to be the cause. Whenever the sublimed
moisture is less than a certain amount, this the sun overpowers
and absorbs : and so causes the wind to cease : but when it is
in excess, it makes the movement of the wind more violent by
the addition of its own impulse.

Sometimes also at sunset the sun makes the wind to cease, by 16
withdrawing the repulsive motion which it gave earlier [in the
day]. And it is clear that this motion has a due proportion [to
the force of the sun] ; so that on the one hand it does not
become spent too soon ; and on the other hand the wind is not
kept longer in motion by it [than while the sun is above the
horizon].

But there is nothing to prevent some winds from blowing
even more as the sun goes down ; such for instance as those
winds which are restrained by heat, and as it were dried up and
burnt up by it.[21] And so at noon, for the most part, [the
movements of such winds][22] lose their force ; but gather
strength as the sun goes down.

The moon produces the same results ; but not in the same 17
degree ; for it is as it were a sun of low power ; for which
reason also the endings and beginnings of the lunar months are

---

[20] See Introd., p. 14.
[21] Cf. the seaman's expression, "the sun is eating up the wind."
[22] πνοαί, subaud.

more terrible at night, and are more stormy [than other parts of
the month].[23]

Thus then it happens that winds sometimes rise, sometimes
fall, at sunrise; and similarly at sunset; for sometimes it stops
them altogether; and sometimes it, as it were, lets them loose.
And it would be well to consider whether these phenomena
happen according to a regular concurrence,[24] as is the case with
phenomena which are to be observed at the risings and settings
of the stars.

18   And from the same cause it is that calms occur very frequently both at midnight and at noon; for if it happens that
the air under such conditions at one time conquers, and at
another is conquered by, the sun. At midnight it conquers,
because the sun is then most distant; at noon it is conquered;
but whether conquering or conquered it comes to a halt; and a
halt is a calm.[25]

It is a fact also that the cessations of winds occur according
to a law. For the winds begin to blow either at dawn or at
sunset; and those which begin from dawn cease whenever
they are conquered; and they are conquered about mid-day;
but those which begin from sunset cease whenever the sun
ceases to have power; and that happens at midnight.

19   Now if some marvel, as at an inexplicable fact, that winds,

---

[23] Cf. Weather signs, § 5, infra., p. 54.

[24] The words κατὰ σύμπτωμα, which I have translated "according to a
regular concurrence," are translated by Schneider and others by "fortuito,"
i.e., "by chance"; a meaning which no doubt the words will bear, but which
the context shows to be here inapplicable. The risings and settings of stars
do not happen "by chance," but synchronize with other events. Our author,
far from suggesting that the varying phenomena he has described are the
product of chance, here proposes as a matter for future investigation whether
they do or not concur or synchronize with other phenomena. The exact
equivalent of σύμπτωμα is "coincidence," which primarily means "happening together." The idea of chance now usually connected with that word is
due to a modern and secondary meaning which it has acquired.

[25] The language throughout is that of military metaphor which I have
sought to preserve.

although they are caused by the impulse of the sun, and, in a word, by heat, should be cold, that which seems to them inexplicable is not a fact at all. For the wind is to be attributed not to the sun simply, but to the sun as, as it were, a co-ordinate cause; nor is it true that movement produced by heat is in all circumstances accompanied by heat and fiery; but [it is so] if it occurs in a particular way. For when a discharge is in the mass and in immediate contact with the discharging agent, it is hot. But when it takes place little by little, and through some narrow channel, then although it is hot in itself, yet the air set in motion by it renders the discharge such [as regards temperature] as the air itself may happen to have been in the first instance.

The breath from the mouth is a sufficient example of this. 20 Some say it is both hot and cold; but they are wrong. It is always warm; the difference is in the manner of emission and escape. For when the mouth is agape, and the breath is sent out in a volume, it is warm; but if it comes out through a narrow opening with more force, and repels the air next to it, and that repels the air next beyond, then if these latter are cold, the current and motion becomes cold also. The same thing happens also in the case of the winds. For when the first movement is through a narrow space, the wind itself is not cold at first; but whatever may happen to be the condition, as regards heat and cold, of the air set in motion by it, such it becomes; if there was heat, it becomes hot; if there was cold, it becomes cold. For this reason winds are warm in Summer but cold in Winter. For according to the particular season such is the air.[26]

---

[26] I fear it must be admitted that the argument of this and the preceding section is almost puerile, and is certainly fallacious. There is obviously no analogy between the difference in temperature of the breath, in the one case

21   It is clear where by reason of the . . . . it happens to be
as it were burnt up. For if where . . . . wind and desire
. . . . hot or cold nevertheless . . . . difference of the
air . . . . such as it may be and it appears to be. . . . . And
in the actual places of its origin and those adjacent, the
current becomes hot ; but as we advance further, it is not so to
the like extent. Sometimes also a wind coming to us from
other parts, if it be from torrid places and such as have a close
and burnt up air, appears excessive in its heat. For which
reason travellers on the road, and men engaged in the harvest,
often die, by reason of such winds, in the fields and in close
suffocating places ; partly from the air which was there before
operating together with the other, and partly from the excess
produced by the current and influx.[23]

22   The following consideration shows that this wind-movement
of the air is not simply due either to its being set in motion by
itself, or to its being forced into motion by heat. If it were set
in motion by itself alone, then as it is cold by nature, and of
the character of vapour, the motion would be downwards ; but
if by heat alone the motion would be upwards ; for that is
the natural direction of fire. But, as a fact, the motion is as it

---

expelled from the lungs when it has been warmed, and in the other merely
taken into the mouth to be at once expelled, and the difference in the temperature of the winds. That wind feels colder when passed through a
narrow opening, producing what we call a draught, is simply due to its
physical effect upon our bodies. It is as a fact not actually colder at all.
And the narrow space or opening which our author requires the wind to
pass through in order to lower its temperature, exists only, of course, in his
imagination, which suggested to him that a wind, starting from some particular point in space or on the earth's surface, necessarily moves at first
within a confined area. The fallacy too is that he fails to give any reason for
the heat or the cold, as the case may be, of the air set in motion by the initial
movement. The premises are so far removed from what we know to be the
facts, that it is difficult to follow the argument or understand the conclusions.

   This passage is hopelessly mutilated and corrupt.

   This passage is incomplete in the text; but the meaning is fairly clear
from what precedes.

were compounded of both; for the reason that neither can [entirely] overcome the other.[29]

To this universally general principle, that such as is the air, 23 or the state of "sublimation" in particular places, such will be the winds in regard to cold, the following facts bear further testimony. All such winds as blow from rivers and lagoons are cold, because of the dampness of the air. For as the sun fails [by its inability to pierce the mist] the air becomes colder and the mist at the same time more dense, particularly if the latter arises close by. So that whenever it strikes the body, a sort of shivering is produced.

On this account also hollow places, and such as are well 24 sheltered from external winds, are chilled by winds arising locally. For the air raised by the sun has neither the natural capacity, nor the power, to remain stationary; and so is borne along and produces a current. And therefore breezes from rivers and lagoons, and generally such as blow off the land, blow at dawn, when the mist is cooling down from the failure of heat. For it is reasonable that this kind of breeze should arise particularly from calm. And they blow still more when there are drizzling rains and moderate showers; for then there is additional matter everywhere for their production; and breezes off the land particularly occur after such conditions.

---

[29] This reasoning is very unsatisfactory. If two forces act in opposite directions, one vertically upwards, and one vertically downwards, the result, if the forces are equal, is equilibrium; if unequal, the resultant is vertically upwards or downwards, according to the direction of the greater force. To produce a horizontal movement two forces must either act both horizontally (whether in the same or in opposite directions), or else must both act obliquely to the horizontal at angles proportionate to the measure of the forces respectively. But the upward force due to heat must be vertically upwards; for there is no reason why it should incline in either direction. And so also the downward force due to cold must be vertically downwards. Therefore no wind movement arising from heat and cold, acting as Theophrastus suggests, could be in a horizontal, or in fact, in any other than a vertical direction.

25   The Nile is the only river from which breezes do not appear
to blow; or if they do, they are very slight, and the reason is
that both the place from which, and the place to which, it
flows are warm; and breezes exist when moisture is condensing.[30]
For which reason also, in no circumstances do breezes come
from any one of the rivers in Libya. For they are warm
throughout, and it is certain that the same is the case with
the rivers about Babylon and Suza, and tropical places
generally. And yet they say that the air becomes marvellously
chilled there towards dawn. This certainly must be enquired
into; for it may be that although the air does become cold, yet
it cannot advance and cause a breeze, because the places which
would at once receive it are hot.[31]

26   The "alternating winds" are produced by the land breeze
and similar breezes: the damp air being gathered together.
For the alternating wind is a sort of reflux of the wind, like the
reflux of the water in tidal straits. For when it is gathered up
and has acquired volume, it changes again to the contrary
direction. These winds occur mostly in valleys, and where
"off-shore" breezes blow. There is good reason for both these
facts. For in valleys the air as it pours in will be collected;
but in open places it is dispersed. Winds too from the land
are naturally weak; so that they cannot force their way far.
And the reflex action is in proportion to the duration and force
with which the winds off the land blow; and it corresponds with
them also as to the time of its occurrence, according as these
winds blow later or earlier.

---

[30] i.e., all parts of the river being equally warm there is no place for
condensation; and consequently no "material" (ὕλη) for the production of
wind.

[31] I suppose that our author means that the heat of the places in immediate
proximity to the river would "restrain and dry up" (see § 16) any breeze
that started from the river; and so in fact it would be imperceptible at any
distance from the river.

There is also a sort of rebound of the winds, so that they 27
blow back against themselves, when they fail to surmount the
places against which they blow by reason of the superior height
of such places. Thus it happens that clouds are sometimes
borne, by an under-current, in directions contrary to the winds.
As for instance in Ægæa of Macedonia, the clouds are carried
towards the North, while the N. wind is blowing. The reason
of this is that, the hills around Olympus and Ossa being high,
the winds fall on and do not surmount them, but are turned
back in the opposite direction; so that the clouds, being at a
lower level, are also carried[32] in the contrary direction. The
same thing happens in other places as well.[33]

Sometimes also, just before the monsoon, contrary winds 23
blow in the opposite direction to the [then prevailing] N. wind
by a reversal of the latter; so that by means of them the ships
make a return voyage; as happens in fact on the passage from
Chalkis to Oropus;[34] and these winds they call "return N.

---

[32] The common reading φέροντι, and Schneider's suggestion of φέρονται,
are equally wrong. It must be φέρεται.

[33] I have frequently observed that when a strong east wind is blowing
across the Severn Valley towards and over the Malvern Hills, the wind is
imperceptible on the eastern slopes themselves, about 900 or 1000 feet below
the summit.

[34] The voyage from Chalkis on the W. coast of Eubœa to Oropus near the
mouth of the Asopus in the N.E. corner of Attica takes a direction nearly
N. and S. down the Euripus. This passage and several other references
indicate a close acquaintance on the part of our Author with the Island of
Eubœa; and it is not uninteresting to remember that Aristotle retired to
Chalkis in B.C. 323 to escape the malice of his detractors at Athens, and there
he shortly afterwards died. We can easily suppose, without undue speculation, that Theophrastus often made the very voyage he here describes from
Oropus (the port nearest to Athens for the purpose) to Chalkis to visit his
great master in his last days, returning again by the aid of the "return north
wind" to supply his place at the Lyceum.

Dr. Donaldson almost scorns to notice "the absurd story" that Aristotle
committed suicide by drowning himself in the Euripus because he could not
discover the cause of the seven tides there; and I only mention it to suggest
that the tidal phenomena in question may have had a connection unsuspected
by Theophrastus with the reversal of the wind. Often in past years have I

winds"; and this especially happens when the winds are very
fresh; for they can make the longest reach at times when the
reverse wind has full power.

Sometimes also it happens that, by striking on a resisting
object, the wind is parted so that it flows hither and thither;
even as water flowing from one and the same opening is divided
by an obstacle into two streams.

29   Altogether there are numerous changes in the winds produced
by local causes; particularly the becoming more violent, or more
calm, according as they blow through a limited, or an open space.
For that which blows through a limited space is always more
violent and fresh; just like a stream of water; for when
collected it has more force and propulsive power; for which
reason, when elsewhere there is a calm, there is always a wind
in narrow gorges; for the air cannot remain there by reason of
its quantity; and the movement of air is wind. So also when
the winds are shut in, or meet in narrow passages and gateways,
they blow with keenness; and windows always draw and produce
a current. Of all these and such like phenomena there is one
and the same cause, namely that which has been mentioned.

30   Again, some places by reason of their situation in valleys, and
being surrounded by greater elevations, happen to be entirely
free from winds, although they are near, or nearer than other
places, to the sources of the winds; while those that are further
off are windy; as happens in Thessaly and Macedonia at the
time of the monsoon. For it does not blow at all, practically,
in these parts; but it blows freshly enough in the far distant
Islands. And the reason is that the former places lie in valleys
and are well sheltered; but the Islands have nothing to oppose

---

watched at Chepstow the cessation of a storm, the clearing of the weather, or
the shift of the wind coincidently with the beginning of the ebb of the great
(though generally misrepresented and exaggerated) tides in the Wye.

the current of the wind. And the monsoon, and generally every wind, is prevented from blowing either by distance (for it cannot extend further by reason of the length of the course), or by the interposition of some objects; or thirdly, if a wind of local origin blows the other way with more force.

Now that the rising of the monsoon, and the blowing of the 31 alternating winds over Macedonia, occur simultaneously at a particular time, must be considered as due to some connecting cause.[35] For the winds everywhere cease at mid-day by reason of the sun; but rise again as the afternoon comes on. And it happens that both the rising of the alternating winds against the off-the-land winds, and the rising again of the monsoon take place at the same period. For I suppose we must not credit the rebound of the wind from Olympus and Ossa with causing the monsoon unless[36] . . . . or very moderate. However we must endeavour accurately to ascertain in all cases the connecting causes of concurrent phenomena.

There is another matter which might appear strange and 32 unaccountable; that is, why it is that among elevated places, those which face a particular wind do not experience that wind at all; but those which are sheltered from a particular wind do feel it, and that not to a moderate extent but severely. For instance, Platæa of Bœotia lies towards the North, and there the N. wind is but a light air; but the S. wind is strong and stormy, although Kithæron[37] stands as a barrier before the place. Again, before the monsoon the alternating winds pass by the low lying lands of Eubœa; but at Karystos[38] they blow

---

[35] See note on § 17.

[36] This passage is mutilated.

[37] Kithæron is the W. part of the lofty range that separates Bœotia from Megaris and Attica.

[38] Karystos lies almost at the extreme S. of the Eubœa protected by hills on the N. and E., and is elsewhere open to the sea.

33 in such a way, that their force is extraordinary. Once more, in that part of Kurias[a] which is called Phæstum, which lies towards the South and is high and precipitous, a marvellous wave breaks in from the sea; but there is no wind; but the ships even are anchored to the . . . . ; the parts in the neighbourhood having no harbour; and there they can watch events. And the reason why the wind does not reach the land is that the air [on the land] does not give place, or flow away by reason of the height towards . . . . not surmounting.[10] But that the air must always move away, and must not remain stationary [in order to admit of a wind blowing] is plain. For in rooms whenever one closes the [inner] doors, the current of air through the windows is reduced; for the room being full of air, and not providing an exit [for the air already within it], does not allow the outer air to enter. For the movement of air is towards a void; and for this reason the expression "draught"[11] is not well used.

34 But the reason why places sheltered from the N. wind (or generally from any particular wind),[12] feel [that] wind more [than places not so sheltered] is that the wind is as it were piled up, and at last overflows, and falls on the place in a mass; for whenever it falls it comes down in a mass like a cataract. And it is in such places that squalls occur; for here are swirlings and massing together; so that when it bursts forth it comes down as it were with a blow. For when the wind is

---

The most southern promontory of Cyprus.

This passage again is mutilated; but the sense is fairly clear. A wind is supposed to be blowing out at sea, causing the swell on the shore. Above the shore rises a mountain, which prevents the air on and about the coast moving away so as to admit of the wind on the sea coming in to the shore. The mutilated sentence probably explained that the wind might rise as it approached the shore, and surmount the obstruction formed by the mountain and the mass of inert air at rest in front of it.

Lit. "drawing." Equally inaccurate is our other expression, "suction."

I have here preferred Wimmer's reading to Schneider's.

massed together, it is violent and non-intermittent; as also is
the case with whirlwinds.

Such and the like are the occurrences due to local causes.
But there are many things which occur in many different
places, to speak of which separately would require a volume.

The following phenomena have relation to all winds in com- 35
mon, being such as in each case afford indications when the
wind is about to blow.

The air varying in its opacity in proportion to its density or
rarity, or in proportion to its heat or cold, or in proportion to
some other condition, always indicates the coming current.
For the conditions of the air sympathise with the movements of
the wind, and precede the winds in affecting our senses.

So also in regard to the sea and waters, it is possible to
observe the same indications; since the waves lifting and
breaking [before the wind comes] indicate that winds are
coming. But they are propelled not without intermission, but
at intervals:[13] and one wave propelled by the wind propels
another: and is again propelled by another puff of wind, when
the first had died away; and so being thus constantly propelled
they arrive at the shore. But when that which is set in motion
has arrived, it is clear that that which set it in motion will
presently come.

It also happens that the waves continue after the winds have
ceased; for they die down and fade away later; because that
which is more difficult to set in motion[14] is also more difficult to
bring to a state of rest.

The following also are common indications of the majority 36
of winds; such as the appearance of shooting stars, and the

---

[13] This seems to mean that the waves come in as long rollers, and not as
continuous breakers.

[14] The sense requires δυσκινητότερον instead of the common reading
δυσκινητότατον.

appearance, fading away, and breaking up of parhelia, and other such phenomena. For the upper air manifests, by the manner in which it is affected, the propagation of the wind before [it is perceived below].

Again, the blowing with greatest force at the end (which is a common feature of most winds) [is an indication to be observed]. For when they blow as it were in a mass, there is little left to come.

Such then are what have been called common and essential characteristics of the winds.

37　Each wind has its own peculiarities, corresponding to its particular nature and position;⁰ some of which are attributable to the places across which, and to which, their currents are directed; others to their originating causes : and others to other such reasons.

The most striking peculiarities in fact are those of the E.N.E. wind (Kaikias) and the W. wind (Zephyros). For the E.N.E. wind (Kaikias)¹⁶ alone attracts the clouds towards itself as the proverb says :—

"To himself he gathers alway, as doth Kaikias the clouds."

38　The W. wind (Zephyros) is the most gentle of all the winds ; and it blows in the afternoon and towards the land, and is cold ; and it blows in two seasons of the year only, namely Spring and late Autumn. There are places, however, where it blows with storm force ; whence the Poet called it "ill-blowing."¹⁷ But in some places it is moderate and soft ; whence Philoxenus in his poems spoke of its "sweet breath." Some fruits also it brings to maturity ; others it thoroughly spoils.

---

¹⁵ That is, in effect, relatively to the meridian or cardinal points. As to the position (θέσις) of the several winds, see Appendix, p. 79.

¹⁶ In these places the MSS. have ἀπαρκτίας. The mistake is obvious ; and I have followed the corrected reading.

¹⁷ Cf. Iliad, xxiii., 200 ; Odyssey, v., 295 ; xii., 289.

The reason [for the peculiarity just mentioned] in the case of **39**
the E.N.E. wind (Kaikias) is, that it is its nature to move in a
curved line, of which the concave side is towards the sky, and
not extended over the earth, as in the case of other winds ; for
this wind blows from below ; and blowing in this way towards
its commencement, it attracts the clouds towards itself. For
towards whatever point the current is, thence also is the
movement of the clouds.[48]

The W. wind (Zephyros) is cold because it blows from the **40**
West, and from the sea and open plains; and still more so
because it blows just after the winter, in spring, when the
sun is only just acquiring power; and in late autumn, when
the sun has no longer power. But it is less cold than the N.
wind because it blows from water in the state of being converted
into wind;[49] and not from snow in the like state. And it is
intermittent because the wind as it is produced is not under
control. For it does not [wait] as winds do on the land [to
gather substance],[50] but wanders hither and thither because it
has come upon[51] a moist surface. And it is uniform and soft
also for this reason ; for it does not blow off hills, nor from **41**
snow rapidly thawed; but it blows easily like [water] flowing
through a pipe.[52] For the regions of the N. wind and the S. wind
are mountainous ; but to the West there is neither hill nor land,
but the Atlantic Ocean; so that it is borne on to [and not from

[48] I must admit that I have failed to make sense of this paragraph. I have
therefore given a literal translation which agrees with the equally literal
Latin translations of Schneider and others. Perhaps the real solution is that
Theophrastus was trying to explain what he did not understand; with the
usual result.
[49] See Introd., p. 15.
[50] The passage is mutilated; but this is obviously the meaning. Cf. § 24.
[51] Or "from." The readings differ.
[52] The readings here are very uncertain. I have done the best I can with
the passage.

or over] the land. And it blows in the evening, by reason of its
place of origin; for all winds are produced concurrently with
the sun diffusing or sublimating moisture through, or into the air,
or co-operating towards their initiation.[52] Whenever therefore
the sun arrives at that particular quarter, then also does the
current from that quarter begin to flow; and this same W. wind

42 (Zephyros) ceases at night because the motive force of the sun
then fails. And it brings the greatest clouds, because it blows
from the ocean and along the sea, so that it collects them from a
great space. And it is stormy and " ill-blowing " for the reasons
already explained; for it blows after winter while the air is still
cold. On the other hand the W. wind (Zephyros) of the late
autumn is not of that character: unless we predicate storminess
of a wind according to its force; for it does blow with force in
some places quite near to others where it has not such force;
just as other winds do. And perhaps its stormy character must
be understood in this way;[51] and not as being general; unless
indeed it be that some writers interchange the names, and call
what is really Thraskias (or the N.N.W. wind) Zephyros (or
the W. wind). But this must be further investigated.

43 Its evenness and softness, where these are present, cause a
certain grateful sensation as it moves and passes by; so that
when this is its character, it is pleasant.

But as to its destroying some fruits, and making others
thrive; that is true of it as a general proposition; and the same
thing can be said generally of other winds as well: for a wind
makes fruit thrive in Summer when it blows cool, and destroys
it when its blast is hot. And similarly again in Winter and
Spring, when the wind is cold, it destroys; but when warm, it
nurtures; and so in each case, it exerts a preserving power by

---

[52] Cf. § 19.

[51] i.e. as due to local causes, and not being universally exhibited.

having a condition of air opposed to that of the season. This happens when the wind is from the sea. For the sea is warm in Winter, and cold in Summer;[53] and for this reason it is that the S. wind has this character in some places; as in Argos; and the N. wind likewise in other places.

What we have now said of this W. wind is its ordinary and common character. But the peculiarities which it exhibits in particular places must be examined; and examined from the point of view of the particular situation, and other circumstances of the place. For the difference will be found to arise almost entirely from local causes. For instance in Italy, Locris[56] and the adjoining country prospers under the W. wind; because it strikes upon it from the sea. But there is another part which does not do so well; and some places are even blasted by it. And again, in Crete, Gortyna thrives under it.[57] For it lies spread out, and the wind strikes upon it from the sea; but another district of that Island on which the wind strikes from off the land and off certain hills, is destroyed. And in the Maliac Gulf[58] it destroys all the seedling crops and the tree fruit; and so too around the Pierian District of Thessaly. The natural

44

45

---

[53] This of course is not true absolutely, but relatively only.

[56] Otherwise called Locri Epizephyrii, which was supposed to be so called by some because it was founded by Locrians from Greece, who then went to "the region of the W. wind"; by others because it was near the promontory Zephyrium. But how this promontory got its name, or how our author conceived that the W. wind struck upon Locri, it is difficult to imagine, for both the promontory and the town lie on that part of the coast of Bruttium (the "toe" of Italy), which faces S.E. towards the Ionian sea; and the W. wind can reach it only from the Tyrrhenian sea, after crossing Bruttium and the range of hills that forms its backbone.

[57] Gortyna lies about the middle of Crete, in the broad valley of the Lethaeus, which runs due W. to the sea.

[58] The Maliac Gulf (now the Gulf of Zeitoun) forms an inlet from the northern end of the strait that separates Euboea from the mainland. Œta lies due S. of it and is separated from it by the pass of Thermopylae. The part of Thessalia called Pieria lies open to the sea on the western shore of the Thermaic Gulf.

configuration of both these places is the same, and the surroundings are similar: for both lie towards the East, and are surrounded by lofty hills: the one by Œta and the hills connected therewith; the other by Pierus. So the W. wind, blowing from the quarter where the sun sets at the equinox,[34] deflects the warmth which strikes from the sun on the hills, and turns it down directly on to the plain, and burns it up. And it acts similarly in other places in which such or the like circumstances occur: and the contrary happens in converse

46 cases. For this, which we repeat again and again, is true absolutely; that it makes a great difference, and especially in respect to heat and cold, through what, and from what quarter, a wind blows. For instance, the reason why the S. wind is not less cold than the N. wind, as the saying is,[35] is that the wind passing through air left as yet chilled and damp by the winter, must, when it strikes upon us, have the same character as the air itself. And the "N.-wind-after-mud" which makes a storm, as another proverb tells us, does so for the same reason: for the air when rendered moist is cold.

So also are breezes from rivers cold as has been already mentioned.

47 The peculiar features of these winds can thus be rationally explained.

But that winds blow in Winter and in the morning from the East, and in Summer and in the afternoon from the West is to be explained on the following ground. When the sun attracting the air can no longer control it, then the air is released and flows along: and so, as it sets, it leaves clouds behind, whence

---

[34] That is to say, from due West.

[35] The saying appears to be lost. It seems from what follows that it was limited to the case of the S. wind in spring. The point seems to be a different one from that mentioned in § 3.

flow the W. winds; and whatever air it draws with it, becomes, to those whose live in the lower hemisphere,[61] a morning[62] wind; and conversely when it sets in the lower hemisphere, it causes W. winds there; but to those here, a morning wind from the air which follows on with it.

For this reason also, if the morning wind find another wind 48 blowing; it becomes greater; for it adds to it.

And as the W. wind is always, and over a considerable area, blowing with those who dwell in the West, so do other corresponding winds blow with those on the other side of the world near our dawn, which is their sunset. These results in fact happen under similar conditions to the inhabitants of each hemisphere, and the wind-current, just as the rain and other phenomena at the extremities of each area, arises according to circumstances; not indeed by any absolute law, but as a general rule.

The S. wind is accustomed to blow at the rising of Sirius, just like any other periodic event. And the reason is that the lower part of the atmosphere is heated by the presence of the sun; so that much vapour is produced. And these periodic S. winds would blow considerably, if they were not prevented by the monsoon from so doing: but, as it is, the monsoon prevents them.

N. winds arising in the night blow themselves out at the 49 third day; whence the proverb runs:—

> " A North-wind rising in the night
> Never sees the third day's light ; "

---

[61] The " lower hemisphere " does not here mean that below or to the south of the equator; but that hemisphere on which the sun shines between our sunset and our sunrise.

[62] The Greek word signifies both " of the morning " and " from the dawn or east." Either sense is here applicable, as the morning wind is from the east.

because the winds that begin from the North in the night are
weak; for it is apparent that the amount set in motion when it
blows at a time when the heat is but small, cannot be much;
for little moves little; and they all end in three days; and those
that are of the least force end early on the third day.

That this same result does not happen also when the S. wind
blows as a night wind, is to be accounted for by the fact that
the sun is near the region that lies towards the South, and the
nights are warmer there, than the days are towards the North;
and the amount of air set in motion during the night is great:
in fact not less so than in the daytime. While the hotter the
days are, the more do they, by drying up the moisture, prevent
the S. winds blowing.

50   And it may be that the reason in the case of the N. wind is
that it bursts all of a sudden as squalls do; and sudden winds
quickly cease:

"For from a weak beginning no great end can come."[63]

The N. wind is also as a general rule violent; and so is the
S. wind after snow and hoar frost, whence the proverb:

"After Frost hoar
Southern winds roar;"

because both [snow and hoar frost] fall when a sort of fermentation and purification has set in; and after fermentation and
purification there is a change to a contrary condition of things:
and the S. wind is contrary to the N. wind. With this agrees
the fact that after rain and hail and such like falls, winds drop;
for all these, and such as these, are a kind of fermentation and
purification of the air.

51   But as cloudy weather, or clear weather, accompanies each

---

[63] I think that the common reading is a corruption of the following Iambic
verse:—

ἀπ' ἀσθενοῖς γὰρ οὐδέν ἐστ' ἀρχῆς μεγά.

wind, according to the country whence it blows and to local causes, so there are some proverbial sayings which relate to certain places only; as for instance that about the W.N.W. wind (Argestes) and the W.S.W. wind (Lips), which they use mostly about Crete and Rhodes;

" Lips is the wind that makes quickly the clouds, and
quickly the sunshine;
" Cloud follows Argestes, all the way unto its end."[64]

For in the places just mentioned the W.S.W. wind quickly produces either result, according to the state of things which exists when it begins to blow; and the W.N.W. wind quickly overcasts the sky.

In some places also there is a sort of sequence of the winds; 52 so that one blows after the other, if the first continue a certain time. And perhaps it is not very strange that, granted the circulation of the winds is always towards those next to them in order, there should on the other hand be also a change over to those of the opposite direction. For there are these two kinds of a change; one when the winds shift round; the other when the winds [that are blowing at first] blow themselves completely out [and others arise]. Of these, the variation by shift is when the E.S.E. wind (Euros) shifts to those next in order of place;[65] and this variation is the less in degree; and when it happens, there is frequently a recurrence to the same point as before, when a storm causes an uncertainty of direction. The variation by changing over is when the wind flies round to the opposite quarter.

This naturally happens in the case of all winds; and in these 53

---

[64] The common reading, ἀργέστῃ δ'ἀνέμῳ πᾶσ' ἔπεται νεφέλη, would mean "every cloud follows Argestes." I have ventured to read πάνθ' for πᾶσ', and to translate it as in the text.

[65] i.e., veers to the S.S.E. (Phœnikias) or backs to E. (Apeliotes).

cases the compensation and reflux, as it were, is such as we
should expect; an instance of which exists in the "off-shore"
winds as the counterpart of the "alternating" winds;[66] and
this order of variation is, in many places, of practically daily
occurrence.

But in some places the counterpart is not an "alternating"
wind, but some second wind from the sea as happens in the
Gulf of Pamphylia. There, in the morning, a wind called
"Dyris" blows with much force, from the river Idyrus; and it
is followed by the S. wind (Notos), and the E.S.E. wind (Euros),
and when they beat against each other mighty waves arise; the
sea is dashed together: many a flash of lightning falls; and the
ships are wrecked. For in every case whenever such a conflict
of winds happens, mighty waves arise, and there is a great storm;
as when, with contrary winds blowing, they say "There is a
battle of the winds." In fact it is but likely that, whenever they
attack each other before they have blown themselves out, it
should cause a storm; for the one adds, as it were, substance to
the other.

This is more particularly evident in the case of the N. wind;
for this wind is of a more stormy nature, and immediately
appropriates the substance that is brought in its way [by another
wind].[67] And in the same way the S. wind is wont to saturate,
and make rainy, any other wind that it conflicts with.

In some places also the S. wind seems to cause snow storms;
as is the case in the neighbourhood of Pontus and the Hellespont,

66 Cf. § 26.

67 This and the preceding passage, as much as, or more than, any other in
the whole book, illustrate the Aristotelian notion of a wind. It is not a mere
force exercised by matter in motion, but an entity existing independently of
matter though requiring matter or "substance," in order to the exhibition of
its power or other attributes; not a condition of matter, but capable itself of
being conditioned, and even capable of appropriating the conditions of its
fellows. See further on this point, Introd., p. 14.

whenever the N. wind has been so cold that it continues freezing the moisture brought up by the S. wind; at least it more frequently freezes than thaws.[67]

The foregoing are what may be called the winter successions 55 and oppositions of the winds.

But the confusion of winds that happens at the rising and setting of Orion happens because, at times of change, everything is naturally liable to get into confusion.[69] Now, Orion rises at the beginning of Autumn and sets at the beginning of winter; so that as there is no established season for the time being, one in fact commencing and the other ending, the winds are of necessity uncertain and confused, because they stand on, as it were, debatable ground between the two seasons. And so it is that this constellation has acquired the reputation of being fierce, both when setting and when rising, by reason of the indefiniteness of the season; for it needs be that it should be disturbed and irregular.

Such, and such as these, then are the phenomena that occur 56 in the air and throughout the Heavens; others are connected with our own conditions. For instance, with Southerly winds men find themselves more weary and incapable; and the reason is that, instead of a little, a great deal of moisture is produced, being melted out by the heat; and so instead of a light air, there is a heavy damp. Again power and strength reside in the joints; and these are relaxed by Southerly winds. For the lubricating matter in the joints when congealed prevents our moving ourselves; but when too fluid[70] prevents our exerting

---

[67] I cannot say that I am satisfied with this; nor are the Latin translations satisfactory. They merely follow word for word the Greek, which I suspect is here corrupt.

[69] The argument here seems to have no foundation, but is an illustration drawn from " la politique."

[70] Wimmer has ὑγρὸν δὲ λίαν οὐ. The latter word should obviously be ὄν.

ourselves. Northerly winds will produce a certain balance, so
that we are stronger and can exert ourselves more.

57　　Again, Southerly winds, when dry and not rainy, produce
fevers; for being naturally warm and moist they induce in our
bodies a warm moisture that is foreign to them; and such a
condition is feverish; for fever is due to the excess of both
these two conditions.[51] But when these winds are accompanied
by rain, the rain cools the system.

In the same way, whatever else affects the habits of our
bodies depends on one or other of these conditions; and such
things are very numerous, and are observed in numerous persons;
but the causes of all are the same, or very nearly so.

58　　So, too, in the case of fruits, and other such like things; for
all the effects which they exhibit are to be referred to either
moisture and diffusion, or density and consolidation, and other
conditions of one category or the other.

So, too, in the case of inanimate things; such as the breaking
of lyre-strings, the cracking open[52] of glued articles, and other
occurrences which happen as things become moist and slack.
For instance, in the manufacture of iron they say that they
can beat it out further with a Southerly wind than with a
Northerly; and the reason is that Northerly winds dry up and
make hard, but Southerly winds moisten and soften; and
everything is easier to work when it is softened, than when it
has become somewhat hardened. At the same time, however,
[the smiths] are stronger and more active in Northerly winds.

59　　As a general proposition the causes of such phenomena as
these are quite evident, for the consequence follows rationally
from the active principles. But there is sometimes, in the case
of either wind, matter for doubt and enquiry; for instance, if

---

[51] i.e., warmth and damp.
[52] The Greek word indicates the noise made by the opening of the joints;
not the opening itself.

neither hardness nor dryness nor recurrence is exhibited with northerly winds, but the opposite conditions appear: and similarly in the case of the S. wind. For that which is contrary to reason requires a cause to be shown for it; but men accept what is reasonable without a cause being shown for it; for they are clever at supplying what is wanting.

But that winds, when they are cold, dry up moisture more 60 quickly even than the sun when it is hot, and that the coldest winds do so most of all, must be understood to be due to this cause, namely, that they produce vapour, and carry it off as they produce it ; and the colder winds do so more than the less cold ; while the sun produces it and leaves it when it is produced.

Why can it be that it is said :—

"Fear not as much a cloud from the land as from ocean
    in Winter ;
But in the Summer a cloud from a darkling coast
    is a warning" ?

Can it be because in winter the sea is warmer than the land, so that if a cloud is formed over it, its formation is obviously due to a powerful active principle ? For otherwise[73] it would have been dissolved by the air by reason of the warmth of its situation; while in Summer the sea is cold and so are the winds from the sea ; and the land is warm ; so that if a cloud is borne from the land seawards its formation must be due to some active principle more powerful than usual ; for the cloud would have been dissolved, if the active principles had been weak.[74]

That the S. wind does not blow freshly in Egypt for the 61 distance of a day and night's journey from the coast, is utterly untrue.[75] But it is said that the N. wind and W.N.W.

---

[73] That is if such a principle were not at work.
[74] The sense requires ἀσθενής instead of the common reading ἀσθ.ιές.
[75] Cf. § 8.

wind most of all the winds there cover the sky with clouds, and the S. wind bears them away; that winds in the dawn bring clouds, and overcast the sky until the sun has risen; but that it does not rain, because the clouds have no place on which to settle; that the S. wind and the S.E. wind and the other winds from the Southern quarter begin to blow at sunrise, and follow round with the sun; but that the N. wind and the W.N.W. wind begin at sunset, and travel round towards the sun-rising.

62   In Sicily they call "Apeliotes" (the E. wind) what we call "Kaikias" (the E.N.E. wind); some, however, think that it is not the same wind, but a different one; because the one overcasts the sky, and the other does not. Some indeed call "Argestes" "Olympias"; others "Skiron" and the Silicians call it "Derkias"; and some call "Apeliotes" (the E. wind) "Hellespontias"; the Phœnicians call it "Karba"; and those in Pontus call it "Berekyntias."[56]

---

[56] There is some confusion in this last section or the text is corrupt. There are obviously some words missing in the original. It seems impossible that the dwellers in Pontus, at the S.E. corner of the Pontus Euxinus, should name the East wind from Berekyntos the mountain range in Phrygia which lies S.W. of Pontus. Skiron was the name of the rocky coast on the confines of Megaris and Attica, about W.N.W. of the Piræus; and so this was probably the name for the W.N.W. wind among the sailors in the Piræus and on the East Coast of Salamis. The same wind would be called "Olympias," and the E. wind would be called Hellespontias by the sailors in the Thermaic Gulf and the north part of the .Egæan Sea.

ON THE SIGNS OF RAIN, WINDS, STORMS, AND FAIR WEATHER.

———◦⟨✧⟩◦———

WE have in the following pages described, as far as was *1* attainable, the signs of rain, winds, storms and fair weather; some from our own previous observations, and the rest upon information from other persons of admitted authority.

Now, such signs as occur at the risings and settings of stars we must take upon the information of Astronomers.

Such settings are of two kinds; for the disappearance of a *2* star is its setting; and this occurs when the star sets together with the sun, and also when it sets as the sun rises.[1]

In like manner risings are of two kinds: some in the morning, when the star rises before the sun; and others at nightfall, when the star rises as the sun goes down. Indeed, what are called the risings of Arcturus occur in both ways; for in winter it rises at nightfall; but in late autumn in the morning. But, of the other stars which have received names, the majority have their risings in the morning, such as the Pleiades, Orion, and Sirius.

Of the remaining weather signs, some are peculiar to all *3*

---

[1] By "setting together with the sun" is probably meant setting in the west in the same course and in the same way as the sun sets; for if the star sets at the same time as the sun it is not visible. Setting as the sun rises may mean either an actual setting below the horizon, or the disappearance of the star as the dawn overpowers its light. So, in the following passage the star rising as the sun goes down may be either its actual appearance above the horizon, or its becoming visible as the sunlight fades, having, itself, risen above the horizon some time before sunset. Both passages together merely refer to the phenomena of what we call morning and evening stars. They have nothing to do with the annual " risings " of constellations, such as the Pleiades or Sirius. See Introd., p. 11.

places in which there are high mountains and ravines; particularly such mountains as extend from a high elevation down to the sea; for, when winds are beginning to blow, the clouds strike on such places; but as the winds change to opposite quarters, the clouds correspondingly change their position,* and becoming moister settle down by force of gravity into the hollows.

For this reason it is necessary for the observer to consider carefully his situation; for it is always possible to find some such indications as these; and the signs thence derived are the 4 most to be relied upon. For the like reason some persons have become good Astronomers in particular places. For instance, Matriketas in Methymna made his observations from Lepetymnus; Kleostratos in Tenedos from Ida; and Phaeinus at Athens (whose pupil Meto established the cycle of nineteen years), observed the phenomena of the solstices from Lykabettus; and Phaeinus himself came to reside at Athens; while Meto was an Athenian born. And others have studied astronomy under like circumstances.

5   There are also other signs which are learnt from observing the habits* of domestic, and some other, animals; and the ways in which they are affected; but for the most part signs derived from the sun and moon are the most important.

Now the moon is as it were the sun of the night; for which reason also the ends and beginnings of lunar months are apt to be stormy; * because the light of the moon fails from the fourth day of the waning moon to the fourth day of the new moon.* The obscuration of the moon also occurs in a similar way to an eclipse of the sun.

---

* ἀντιμεθίσταται should obviously be read instead of ἀντιμεθίστανται.
I read, with Schneider, τρόπον for τόπον.
Cf. Arist. de Gen. Anim., ii., 1, 9; and On Winds § 17 supra, p. 30.
i.e., three days before and three days after new moon.

He, then, who wishes to forecast, must pay special attention **6** to the rising and settings of these bodies; [and observe] in what circumstances they occur. And, first of all, it must be understood that all measures of time[a] are divisible naturally into two parts; so that in reference to such divisions we must consider the year, the month, and the day. The Pleiades divide the year by their rising and setting; for from the setting to the rising is half a year; so that the whole period is divisible into **7** two halves. And the solstices and the equinoxes have the same operation. Whatever, then, be the condition of the air at the setting of the Pleiades, such it continues for the most part until the [winter] solstice; and if it change, it changes immediately after the solstice; but if it does not change, it continues so until the vernal equinox; and thereafter in like manner until the [rising of the] Pleiades; and from that until the Summer solstice; and thence until the [Autumnal] equinox; and from that equinox to the setting of the Pleiades. And each month **8** follows a similar rule; for the full moons, and the quarters,[b] and the fourth days,[c] divide [the periods] equally; so that we must begin our review from the new moon as a starting point. The change takes place for the most part on the fourth day; and, if not then, on the first quarter; and, if not then, at the full; and from the full moon [it continues] till the last quarter; and thence to the fourth day [of the waning moon]; and thence to the new moon. And the diurnal changes, for the most part, **9** occur according to the same law. For the sunrise, the forenoon, noon, afternoon, the sunset, and the corresponding divisions of

---

[a] ὥραι.

[b] Literally the eighth days, i.e., after change and full, which are with us the first quarter and last quarter.

[c] That is the third day before or after new moon, first quarter, full moon, and last quarter, making, with the day of such occurrence, the fourth day; and commencing or terminating the half of a quarter.

the night produce similar results to those just mentioned, in
relation to winds, storm, and fair weather. For if the weather
is going to change it generally changes at such divisions. In
every case therefore, the measures of time must be taken into
consideration; but, in the case of each sign, in accordance with
the method hereafter stated.

I.

10   Now the signs of Rain are such as the following:—

The plainest sign is that which is to be observed in the
morning, when, before the sun rises, the sky appears reddened
over; and it indicates rain, either on the same day, or generally
within three days; and the other signs shew the same; for rain
is indicated, if not sooner, within three days at the most by a
reddened sky at sunset also, but less certainly than when it is
11 seen in the morning. And if, either in winter or spring, the
sun goes down into a thin cloud,[9] it generally indicates rain
within three days; and so also if there are streaks of clouds
from the Southward; but these same appearances from the
North are less certain. And if the sun, as it rises, has a dark
mark[10] on it, and if it rises out of clouds, rain is indicated: and
if, as it is rising, rays stretch upward before it actually rises,
this is a sign alike of rain and of wind. And if, as the sun is
going down, a cloud comes under it so that the rays are thereby
divided, it is a sign of storm. And whenever the sun is fiery
at its rising, or setting, unless the wind rise, it is a sign of rain.
12 The same is indicated by the moon as it rises at the full; but
less by the crescent moon. If it be fiery, it indicates that the

---

[9] The word νεφέλιον here used is simply the diminutive of νέφος "cloud";
and might equally express merely the size of the cloud; but I think it means
here a semi-transparent cloud.
[10] I think σῆμα ought to be read here in place of σημεῖον.

month will be windy; if hazy, that it will be wet. And whatever the crescent moon indicates, it indicates when it is three days old.

If shooting stars are frequent, they are a sign either of rain 13 or wind; and the wind or rain will come from the quarter whence they proceed. And if, while the sun is either rising or setting, numerous rays arise therefrom it is a sign of rain. And when during sunrise the rays retain a colour as if the sun were being eclipsed, it is a sign of rain. And when the clouds are like fleeces of wool,[11] it indicates rain. An unusual number of bubbles on the surface of the rivers indicates great rain.

The colours of the spectrum[12] seen around, or through, the flame of a lamp usually indicate rain from the south. Snuffs 14 on the wicks indicate rain, if the wind be in the South;[13] but they indicate wind also in proportion to their number and size; and if they are small and like millet seeds and bright, they indicate both wind and rain. And when in winter the lamp is separated from the flame by a space (as it were a bubble), it is a sign of rain; and so also if the rays throb upon the lamp, and if sparks are produced.

If birds which do not live on the water wash themselves, it 15 indicates either rain or storms. The toad washing and frogs croaking more than usual indicate rain. If the lizard called the salamander is seen it indicates rain; and so again does the green frog croaking on a tree. Swallows skimming[14] the ponds indicate rain. The ox licking his fore-hoof indicates a storm or rain.

The cormorant crying on a rock which a wave is washing 16

---

[11] Virg. Georg. i., 397 mentions as a sign of fine weather "tenuia nec lanæ per cœlum vellera ferri—Cf. Aristoph. Nubes 343 ἔρια πεταμένα.

[12] Literally "a rainbow."

[13] Cf. Aristoph. Vesp. 262; Virg. Georg. i., 392.

[14] Lit. "striking with their bellies."

over indicates rain; and so if she dives frequently and flies in circles.

If a raven accustomed to utter varied notes, utters two of these quickly and then croaks and flaps its wings, it indicates rain. And so if, when there are showers, it makes many different calls and sitting on an olive tree picks lice from itself. And if, whether during fine weather or rain, it imitates with its voice the dropping of water, it indicates rain.

If ravens or jackdaws fly upwards and scream [15] like hawks, it indicates rain.

If a raven in fair weather utters an unusual cry and croaks, it indicates rain.

17　If a hawk sitting on a tree, then flies within it and picks insects from itself, it indicates rain.

If in summer many birds which usually live on an island appear in flocks [on the mainland], it indicates rain; if the number of them is moderate it will be good for the goats and cattle : [16] if the number is excessively great it indicates severe drought. And generally birds and cocks pecking themselves is a sign of rain ; and so when they imitate the sound of water as if it were raining.

18　If a tame duck going under the eaves shakes out its wings, it indicates rain. And so also if jackdaws and cocks shake out their wings over a pond or the sea like a duck, it indicates rain.

The heron crying early indicates either rain or wind. And if it cries as it flies towards the sea, it is rather an indication of rain than of wind; but it generally indicates wind by its cry.

---

[15] The word used strictly means "act like hawks." Perhaps an imitation of the flight rather than the cry of the hawk is meant.

[16] This probably means that there will be sufficient showers to produce good pasturage.

If a finch in a dwelling house sings in the morning, it **19** indicates rain or a storm.

A jar quite full [of water] emitting sparks all over [when set on the fire] is a sign of rain.[17]

Many centipedes crawling towards a wall indicate rain.

A porpoise frequently diving and coming to the surface near the shore indicates rain or a storm.

If the Lesser Hymettus (which is called Dry) has a small **20** cloud[18] in its hollow, it is a sign of rain; and if the Great Hymettus in summer has white[19] clouds above and on its side, it is a sign of rain; So also if Dry Hymettus has white clouds above and on its side.[20]

If the S.W. wind blow at the time of the equinox, it indicates rain.

Thunder occurring in winter or in the morning indicates **21** [wind] rather [than] rain;[21] but thunder in summer at noon and in the evening is a rainy sign.

---

[17] I have given here Schneider's explanation of this passage, but it is not satisfactory; and I think something is lost. There is nothing in the Greek to justify the insertion of the words "when set on the fire"; and Schneider makes ὕδατος do double duty, first combining it with περίπλεως, and secondly with σημεῖον. But I have nothing better to offer.

[18] See note on § 11, supra.

[19] Wimmer omits λευκάς (white) in his text, but gives no reason. Probably it is omitted by a printer's error; for he has "albas" in his Latin version.

[20] The Hymettus (Greater and Lesser) is about five miles S.E. of Athens, and forms the N.E. end of the range which runs S.W. and N.E. through Southern Attica, and there the Ilissus rises. It is a prominent feature in the landscape from the greater part of Athens. The Greater Hymettus (or Hymettus proper) has an elevation of 3,368 feet (above average sea level at the Piræus), and is separated from the Lesser or Dry Hymettus (2,558 feet), now called Mavro Vunó and forming the southern part of the range, by a depression the elevation of which is 1,479 feet. This depression is the hollow mentioned in the text. For a description of this and the other hills of Attica, and the geology of the district, see *Geologie von Attika*, by Richard Lepsius, Berlin, 1891-3.

[21] The insertion of the words ἄνεμον ἤ (corresponding to the words in brackets) is necessary to the sense.

If lightning is seen in all directions it is a sign of rain or a
storm : and so too if it happens in the evening.

If there is lightning from the S. at daybreak with a S. wind
it indicates either rain or wind.

The W. wind blowing with lightning from the N. indicates
either a storm or rain.

In Summer, lightning in the evening indicates rain immediately, or within three days.

In early Autumn lightning from the N. indicates rain.

22   Whenever Euboea appears girt in the midst by clouds, there
will be rain in a short space.[22]

If the clouds settle down on Pelius, it indicates rain or wind
from the quarter whence the clouds settle.

Whenever there is a rainbow it is a sure sign of rain.
If many occur, it indicates a great deal of rain. And so
also in many cases when a burning sun breaks forth from a
cloud.

If ants on the side of a hollow carry their eggs from the nest
to the high ground it indicates rain; but if they carry them
down, fair weather.

If two Parhelia occur, one towards the South, the other
towards the North, with a halo round the sun, they indicate
rain within a short time.[23]

Dark halos are a sign of rain, particularly those in the afternoon.

23   In the constellation of the Crab are two stars which are
called the "Asses"; in the space between these is the nebula
called the "Manger."[24] If this becomes hazy it is a sign of
rain.

[22] Euboea has a range of lofty hills about its centre part.
[23] Cf. Arist. de Mund., iv., 22.
[24] Or "Praesepe."

If it does not rain at the rising of Sirius or Arcturus, there will generally be rain or wind about the equinox.

The common saying about flies is true; for when they bite vigorously it is a sign of rain.

When the finch sings in the morning, it indicates rain or a storm: but in the afternoon rain.

Whenever a long white cloud envelopes Hymettus[25] down-24 wards from its peaks at night, rain occurs, as a rule, within a few days.

If in Ægina[26] a cloud settles down upon the temple of Zeus Hellenius rain generally occurs.

If there is much rain in the Winter the Spring is generally dry.

If the Winter is dry the Spring is rainy.

Whenever there is much snow, a fruitful season generally follows.[27]

Some say that, if on the coals[28] when burning there appear 25 bright spots of the size of hailstones,[29] it generally portends hail. But if many, as it were, small bright millet seeds appear, then, if the wind is blowing, fair weather is indicated. But if there is no wind, then rain or wind.

It is better both for plants and animals that rain from the North should precede that from the South; but it should be sweet, and not salt to the taste.

Speaking generally a year with the wind from the North

---

[25] Cf. note on § 20.

[26] Ægina lies in the middle of the Saronic Gulf. On a hill on the N.E. part of the Island, about 12 miles from the Piræus and 17 from Athens (and so a conspicuous object from the latter place), stood the temple of Zeus Hellenius or Panhellenius.

[27] Cf. Theoph. de Plantis ii., 2.

[28] i.e., pieces of charcoal.

[29] Lit. "a bright hail."

is better and more healthy than one with wind from the South.

When the ewes or she goats are covered more than once it is a sign of a long winter.

Such are said to be the signs of rain.

II.

26   The signs of wind and airs are such as the following :—

The sun rising fiery red even though it do not shine is a sign of wind.

If the sun appear hollow,[30] it is a sign of wind or rain.

If it appear fiery for several successive days, it indicates drought and wind, both of long duration.

If about sunrise the rays of the sun are parted, some towards the North, some towards the South, the sun itself being between the two sets of rays, it is a sign equally of rain and wind.

27   Black spots on the sun and moon indicate rain; red show wind.

If the crescent moon stands upright with a N. wind blowing, W. winds usually follow, and the month will continue stormy to the end.

Whenever the upper horn of the crescent moon stoops forward, N. winds will prevail during the period of the new moon;[31] but when the lower horn comes forward, S. winds will prevail.   But if it is upright, or only very slightly inclined, it is usually stormy till the fourth day; or if the disc of the moon is plainly visible then until the first quarter.   When hazy it indicates rain; but when fiery, wind.

---

[30] This is the literal meaning of κοῖλος; but I do not understand what is meant. It may refer to the elliptical or flattened appearance which the sun sometimes presents; but the word does not fairly express such a meaning.

[31] That is until first quarter.

Divers and ducks, both wild and tame, indicate rain by 28 diving; but wind by flapping their wings.

Petrels in fine weather indicate that the winds will blow from the quarter towards which they fly.

Sparrows chattering after evening has set in portend either a change of wind or rain in showers.

A heron flying from the sea and crying is a sign of wind. And generally if its cry is loud it shows wind.

A dog rolling on the ground indicates a force of wind. 29

Many spiders webs borne in the air indicate wind or a storm.

Receding of the sea indicates a N. wind; but its influx a S. wind. For if with the wind in the North an influx of the sea occurs, the wind changes to the South; and if with the wind in the South a recession takes place, the wind changes to the North.

The sea surging and the shores resounding and the cliff moaning are signs of wind.

The N. wind is less as it is ceasing; and the S. wind less as it begins to blow.

A parhelion indicates either rain or wind from the quarter towards which it appears.

The 15th day after the Winter Solstice generally has the 30 wind in the South.

When a N. wind is blowing, [the air] dries up everything, but when a S. wind is blowing it moistens everything.

If when the S. wind is blowing any piece of glued furniture makes a noise it indicates a change to the North.[32] If feet swell the change will be to the South; and the same thing is the sign of a hurricane.

Biting the right . . . . .[33]

---

[32] The ordinary text has νότια - " south "; but the sense requires βόρεια— " north."

[33] The MSS. are corrupt here.

The hedgehog is an animal that gives signs. It makes for
itself two entrances to the place where it lives: one towards the
North, the other towards the South. Whichever of these it
closes it shows that the winds will come from that quarter. If
it closes both it shows that there will be a force of wind.

31   If a hill [is covered with clouds][31] towards the North, it
indicates wind.

If on the sea there is on a sudden a lull of the wind, it indicates either a change or a freshening of the wind.

If headlands far out at sea become visible, or several
Islands appear instead of one, it indicates a change to the
Southward.[35]

If the land appears dark from the sea, the wind will be from
the North; if light it will be from the South.

Halos round the Moon are more indicative than those round
the Sun. But in either case, when they are interrupted they
indicate wind; and wind from that side on which the interrup-
tion occurs.

When the sky is clouded over, the wind will come from the
quarter on which there is a lifting [of the clouds].[36]

Clouds without rain in Summer indicate wind.

32   If lightning appears all round, it indicates rain; and from
that side on which it is frequent the wind will rise.

In Summer-time strong winds rise from that quarter from
which thunder and lightning come. If the lightning is strong

---

[31] There is obviously a hiatus in the text which I have attempted to supply.

[35] The ordinary phenomenon of refraction is of course referred to here.

[36] I differ in this passage from both Schneider and Wimmer. The latter
translates it "unde sol cœlo nubilo exortus fuerit inde venti oriuntur."
The former "unde cœlo nubilo sol exortus fuerit, ab ea parte ventus ingruet."
But in the passage there is nothing about the sun. The verb which I have
translated "lifting" is no doubt also used of the sunrise; but the translations
I have rejected would come to this—that whenever it is cloudy at sunrise
the wind must blow more or less from the East; which is absurd. What my
translation expresses is the commonest experience.

and intense, the winds will blow with the greater velocity and strength: but if gentle and of little intensity, they will blow but little.

In Winter and late Autumn the contrary takes place. For, the more intense then is the lightning and thunder, the more do the winds cease. But in Spring I take less account of these same matters as signs, as also in Winter.

If when the S. wind is blowing, there is lightning in the **33** North, it ceases to blow.

A wind rising in the early morning if accompained by lightning,[37] generally ceases on the third day; other [such] winds cease on the fifth, seventh or ninth day. Winds that rise in the afternoon quickly die away.

N. winds generally cease on odd days; S. winds on even days.[38]

Winds rise at the times of the rising of the sun and the moon. If the sun or the moon on its rising[39] cause the wind to drop, it increases afterwards in force.

Winds which begin to blow in the day last longer and have more force than those which begin to blow at night.

If the " Monsoon " have blown longer than usual,[40] and the **34** late Autumn has been windy, the Winter is free from wind. If otherwise, the Winter is otherwise. In whatever direction a cloud stretches out from the peak of a mountain in that direction will the wind blow.

If clouds settle down on the back of a mountain, the wind will blow from behind it also.

---

[37] I prefer Schneider's reading ἕως ἀστράπτωος to Wimmer's ἐὰν ἔωθεν ἀστράπτῃ.

[38] i.e., counting from their rising.

[39] Some would insert μή, making the sense "do not cause," &c.

[40] Cf. Herod., vi., 140; vii., 168. Arist., Problem. xxvi., 2.

E

If Athos[41] is girt with clouds about its middle, it is a sign of S. wind; and generally speaking mountains so begirt indicate S. wind in most cases.

Comets generally indicate wind. If there are many they indicate drought also.

The S. wind generally blows after snow, the N. wind after frost.

Snuffs in the lamp indicate either wind or rain.

35 The directions of the winds are such as have been described in the drawing.[42]

Of all the winds the N. by W., N.N.W. and W.N.W. most usually blow against others while still blowing.

When winds are not neutralised by each other but blow themselves out, they change into the winds next to them on the right hand as the path of the sun goes.

The S. wind when beginning to blow is dry: but at the close is wet; so is the E.S.E.

The E. wind from the Sunrise of the equinox is rainy; but it brings showers and light breezes.[43]

36 The E.N.E. and W.S.W. are chiefly wet; N. by W., N.N.W. and W.N.W. bring hail; N.N.E., N., and N. by W. bring cloud; S., W., and E.S.E. bring heat; some of them to those to whom they come from the sea, others to those to whom they come across the land.

The E.N.E. chiefly, and then the W.S.W. makes the sky dense, and covers it with clouds.

All other winds drive the clouds before them; the E.N.E. alone draws them towards itself.

---

[41] Athos stands 6349 feet above sea level on the promontory of Acrathos in Macedonia.

[42] See Appx., p. 79; and cf. Arist., Met. ii., 6.

[43] For the Greek expression, cf. Eur. Iph. Aul., 813.

The N.N.W. and W.N.W. chiefly produce bright weather;
and of the rest the N. by W.; but the N. by W., N.N.W.
and W.N.W. chiefly produce hurricanes.

Hurricanes[44] occur when winds conflict with each other 37
principally in late Autumn and next in Spring.

The N.N.W., W.N.W., N. by W., and N.N.E. are accompanied by lightning.

If much acanthus down is borne along on the sea, it shows
that there will be a great wind.

When many stars shoot from one quarter, it shows that there
will be wind from that quarter.

If they shoot from all quarters alike, it shows that there will
be winds from many points.

Such are the signs of winds.

III.

The following are the the signs of storm :— 38

The sun setting into a bank of haze;[45] and according as is
the proportion of the disc so covered as it sets such will the
[following] days turn out; for instance, if a third or a half
be obscured.

If the new moon be upright until the fourth day, or the
whole disc be plainly visible, there will be stormy weather until
the first quarter.

If cranes fly early and in numbers there will be an early
storm; but if late and for a long time, the storm will come late
And if they wheel in their flight they indicate a storm.

Geese cackling more than usual or fighting for their food is a 39
sign of storm.

---

[44] The Greek word for hurricane is very expressive; it literally means an
"out-of-a-cloud." Cf. the American expression "cloud-burst."

[45] Lit. "into a not pure space."

The finch or sparrow chirping at dawn is a sign of storm.

The wren going under cover and entering into holes indicates storm: and the redstart likewise.

If the crow calls twice quickly and then a third time, it indicates a storm.

The crow, raven and jackdaw calling late indicate storm.

If a sparrow or swallow or bird of any other of the species that are usually black appear white, it indicates a storm; just as black ones seen in numbers indicate rain.

40   If birds fly in as for safety from the sea they indicate a storm.

A finch singing in a dwelling house indicates storm.

Whatever indicates rain is followed by storm: and if it is not followed by rain it is followed by snow and storm.

If the raven makes several different cries in the Winter it is a sign of storm.

Jackdaws flying from the South and cuttle fish[46] indicate a storm.

A voice re-echoing in a harbour and making a confused sound indicates a storm.

If jelly fish[47] appear in numbers in the sea it is a sign of a stormy year.

Sheep copulating early indicate an early Winter.[48]

41   If in late Autumn sheep or cattle scratch up the ground and lie together in numbers with their heads towards each other it indicates a stormy winter.

---

⁴ I suspect the text, and think that probably the name of some bird has been corrupted.

⁴ I have thus ventured to translate πτέρυον θαλάττιος; the only information obtainable from the lexicographers being that it is "a kind of mollusc"! The word is very descriptive of a jelly fish, and my own observation agrees with the statement in the text. In the fine year 1893 jelly fish were very infrequent on the S.E. coast.

⁴ Cf. Plin. xviii., § 85.

In Pontus they say that when Arcturus has risen, the flocks prefer to feed facing the North wind.[49]

Cattle eating more that usual and lying down on the right side indicate storm.

An ass shaking its ears indicates storm; and so do sheep and herds fighting for their food more than usual. For they are preparing beforehand. Mice squeaking and dancing about indicate a storm.

A dog scratching up the ground with its paws, and the tree **42** frog croaking[50] by itself at daybreak are signs of storm.

The appearance of many earth worms indicates a storm.

If the fire will not light it is a sign of storm; and if a lamp refuses to be lit it indicates a storm.

Ashes binding together indicate snow.

A lamp burning slowly in fair weather indicates a storm; and if in Winter black snuffs collect on it, it indicates a storm; and if it becomes covered over with, as it were, many millet seeds, there will be a storm; and if in fair weather they collect in a circle round the flame, it is a sign of snow.

If the "Asses Manger"[51] is condensed and hazy, it indicates **43** a storm.

If bright lightning does not continue in the same place, it is a sign of storm.

If, at the setting of the Pleiades, lightning is bright over Parnes, Brilettus, and Hymettus,[52] and it shine over all, [at the same time] it indicates a great storm; if on two only then a less one; and if over one only then fair weather.

[49] This passage is corrupt.

[50] Or it may be an owl or thrush singing alone. See Theocr., vii., 139.

[51] Cf. § 23, supra.

[52] This (like many other passages) supposes the observer to be at Athens. Parnes lies 13 miles due N. of Athens and is the eastern (as Kithaeron, see note, p. 37, is the western) point of the range running east and west, which separates Bœotia from Megaris and Attica. Brilettus lies about the same

If during the Winter there is a long cloud over Hymettus it indicates a prolongation of the winter.[53]

Athos, Olympus and the peaks of mountains generally if covered by cloud, indicate a storm.

If in fair weather a thin cloud[54] appears stretched in length and feathery[55] the Winter will not end yet.

44   If the late Autumn is unusually bright the Spring is cold, as a general rule.

If the Winter sets in early it closes early, and the Spring is fair; but if the contrary, the Spring also will be late.

If the Winter is wet the Spring is dry; if the Winter is dry the Spring is fair.

If the early Autumn is mild, the sheep generally suffer from famine.[56]

If the Spring and Summer are dry, the early Autumn, and the late Autumn as well, are close and free from wind.

45   If the Scarlet-oak[57] be full of berries there will be very many storms.

If a cloud stands upright on the peak of a mountain it indicates a storm; whence Archilochus wrote in his poem; "See, Glaucus! the deep sea already is surging with waves; and around the tops of the hills an upright cloud stands encircling them; the sign of a storm."

---

distance to the N.E., separating the Vale of the Cephissus from the Plain of Marathon. This range attains an elevation of 3,696 feet, and is more generally known as Pentelicus; and has numerous marble quarries, which no doubt supplied the builders and sculptors of Athens. The marble beds, like that of Hymettus, are of the Cretaceous period overlaid by tertiary strata. Hymettus (v. supra., § 20 and note) lies five miles to the S.E.

[53] In both places in this passage "storm" may be read for "winter."

[54] Cf. note on § 11.

[55] Lit. "plucked at."

[56] Cf. note on § 17.

[57] This produces the kermes berry, whence the scarlet dye—Κόκκος. Cf. Theophr. de Plantis, iii., 7, 3.

If the cloud be like in colour to a white skin it is a sign of storm.

When clouds are stationary, and others accumulate by them, but the first remain still, it is a sign of storm.

If the sun in Winter shines out, and is again hidden, and this **46** occurs twice or thrice, the day will be stormy as it goes on.

Mercury, when seen in Winter, indicates cold; in Summer, heat.

When bees do not fly afar but fly about in the same place, it indicates that a storm will follow.

The wolf howling indicates a storm within three days.

If the wolf comes hurriedly towards or into the farm in the Winter season it indicates that Winter is at hand.

It is a sign of great storms and showers when there are many **47** wasps in the late Autumn; and when white birds[54] come near to the farms; and generally when wild animals approach the farms it is a sign of Northerly winds and violent storm.

If those parts of Parnes which face the W. wind and the parts about Phyle are covered with cloud when Northerly winds are blowing, it is a sign of storm.[55]

When there is very close hot weather there is generally a **48** re-action and a severe storm follows.

If there is much rain in the Spring, great heat follows in flat places and valleys. The beginning [of the year] must therefore be watched.

If there is a great deal of bright weather in the late Autumn, the Spring generally is cold; but if the Spring is late and cold, the early Autumn is late, and the late Autumn is generally close and hot.

---

[54] Probably sea-fowl.

[55] As to the position of Parnes, see note on § 42 and map B. Phyle lay on the road from Athens to Thebes, which traverses the pass separating Kithæron from Parnes.

49   When the scarlet oaks[60] are very full of berries they
generally indicate a severe Winter; but they say that sometimes drought follows.

If one takes a woodcock and puts it under a wine jar with
clay plastered round the bottom, it indicates by the cries which
it utters, wind and fair weather.

When mice fight for chaff and carry it away, it is a sign of
storm : as is everywhere commonly reported.

IV.

50   The following are the signs of fair weather :—

The sun rising bright, and not fiery, and without any
marking[61] upon it, indicates fair weather.

So also does the moon at the time of full moon.

The sun setting in Winter unobscured is a sign of fair
weather; unless on the preceding days it has set out of a clear
sky behind a bank of haze ; and in that case the forecast is
uncertain.[62]

If during a storm the sun sets unobscured, it is a sign of fair
weather.

If when setting in Winter its colour be pale yellow, it
indicates fair weather.

51   If the crescent moon is bright on the third day, it indicates
fair weather.

---

[60] Cf. § 45.

[61] Cf. §§ 11 and 27. This may refer to spots on the sun ; but it may as well
mean thin dark lines of cloud partially obscuring the disc. For σημείον,
as meaning a device on a shield, see Herod., i., 171 ; Eur. Ph., 143, 1114. The
remarkable spots on the sun recently noticed were clearly visible to the
naked eye at sunset on September 4th, 1895; so it is quite possible that it is
such as these to which Theophrastus is referring.

[62] Cf. supra. § 38.

Whenever the "Asses Manger"[63] is clear and bright, it signifies fair weather.

If a halo gathers and fades uniformly,[64] it indicates fair weather.

Hollow clouds[65] in Winter indicate fair weather.

When Olympus, Athos, and generally all hills that give indications, have their tops clear, it indicates fair weather.

Whenever the clouds girt the mountains quite down to the sea, it is a sign of fair weather.

So whenever, after it has rained towards sunset, the clouds have a colour like copper; for it is fine generally on the next day.

Whenever there is a fog, there is little or no rain. 52

Whenever cranes take flight and do not return, it indicates fair weather. For they do not fly away before they fly about and see that the sky is clear.

An owl hooting quietly in a storm, indicates fair weather; and [also] when it hoots quietly by night in Winter.

The sea owl[66] crying during a storm, indicates fair weather, but crying in fine weather indicates a storm.

A raven by itself croaking quietly, and also if it croak thrice and then several times, indicates fair weather.

The crow, if it caw thrice immediately after daybreak, 53 indicates fair weather, and also when it caws quietly in the evening in the Winter.

The wren flying out from its hole or out of enclosures and out of a house,[67] indicates fair weather.

---

[63] Cf. supra. § 23.

[64] That is, without being broken up or interrupted on one side or the other. Cf. supra § 31.

[65] I cannot attempt an explanation of this term.

[66] I do not know what bird is meant by θαλαττία γλαύξ, and I have merely translated the term literally.

[67] Of course, the difference in structure between a Grecian house and a modern one will not be forgotten.

If during a storm, with the N. wind blowing, a white under light appear from the North, but on the South a cumulus[57] cloud is extended opposite to it, it generally indicates a change to fair weather.

Whenever the N. wind blowing strongly brings up many clouds it is a sign of fair weather.

54   Ewes being covered late [in the season] is a certain sign of fair weather.

An ox resting on his left thigh indicates fair weather, and the dog likewise; but lying on the right indicates a storm.

Many grasshoppers indicate that the year will be pestilential.

A lamp burning quietly in a storm (winter) indicates fair weather. So also if on the top the lamp has as it were bright millet seeds, and if it has a bright line described round the wick.

55   The fruit of the mastick tree foreshows the periods of sowing. It has three divisions; the first fruit is the sign of the first period: and the second of the second; the third of the third; and whichever of these turns out the best and best grown, so will be the corresponding sowing time.

56   The following are said to be signs of entire years and of parts :—

If at the beginning of Winter there is dark weather and heat, and these pass away under the influence of winds without rain, it indicates that hail will follow towards Spring.

If mists occur after the vernal equinox, they indicate airs and winds till the sixth month thereafter.[69]

Mists which occur with the crescent moon indicate winds until that time.[70] But those that occur when the moon is doubly convex indicate rain.

---

[57] Lit. "swollen" or "turgid"; "that onward drags a labouring breast."

[69] Lit. "till the seventh month, counting both."

[70] This probably means until the corresponding period of the next moon.

In proportion as fogs occur with each of these phases of the moon, the more do they give the indications mentioned.

Winds occurring with the happening of mists have their 57 signification; and if the winds come from the East or South they indicate rain; but if from the West or North they indicate wind and cold.

What the Egyptians call comets not only indicate by their appearance what we have already said, but cold also.

It is usual also for there to be indications at the time of the [appearances of] stars and of the equinoxes and solstices; but they usually occur not on the day but shortly before or after.

INDEX OF PLACES MENTIONED.

NOTE.—The Latitudes and Longitudes are taken from the Rev. G. Butler's Public School Atlas of Ancient Geography. The names of Districts and Provinces are printed in Capitals, and of Mountains in Italics.

THE HOROLOGIUM OF ANDRONIKOS.

APPENDIX.

On the Number, Direction and Nomenclature of the Winds
in Classical and Later times.

The idea of the division of the heavens into four quarters,
and of winds blowing from each of those quarters, is one
familiar to students of Biblical and other ancient literature;
and may be said to have found its final expression in the
description of

> "The tower that stood four-square
> To every wind that blew."

Such a division and classification would satisfy the require-
ments of remote ages; but as the ever varying "signs of the
sky and of the earth" claimed man's attention, a more accurate
division and some more definite means of determining the
directions of the winds that preceded or accompanied such signs
became necessary. To trace the development of this division, as
far as we can learn it from classical and post classical authors
and monuments, is the object of this Appendix, with a hope
that it may aid in a better understanding not only of Theophrastus, but of other ancient scientific works.

Homer (before B.C. 800) names four winds only: Boreas,
Euros, Notos, and Zephyros. These, therefore, it seems safe
to infer to have been, in his time, referred to the four cardinal points. But I do not think it is to be assumed that
these four distinct and principal winds were the only winds
then recognised. Indeed in the myth (Il. xx., 223—229)
of the twelve colts begotten by Boreas of the mares of
Erichthonios ("which galloped over the tops of the flowers and
brake them not, and over the crest of the ocean wave"), one

seems to see a reference to the twelve winds of a later era. Still there is no trace at that time of a distinctive name for any wind other than the four.

When Homer speaks of Boreas and Zephyros blowing from Thrace (Il. ix., 5), or setting out together from the home of Zephyros for Troy to fan the funeral pyre of Patroklos (Il. xxiii., 192-218), we are not to suppose that he confuses the two, or indicates that the direction of Boreas is in a line from Thrace to Troy. The mountains of Thrace are the poetic home of the winds; and Theophrastus in many places shows why this is properly so. In the passage last referred to, the morning visit of Iris to Thrace, followed by the blasts of the north and west winds, represents the rainbow of the morning in the western sky as the proverbial precursor of storm.

Homer (as mentioned by Theophrastus, *On Winds* § 38) assigns opposite attributes to Zephyros; representing it as stormy (Il. xxiii., 200; Od. v., 295), rainy (Od. xiv., 458), and soft and gentle (Od. iv., 567); while the violence of Hector's attack on the Greeks is compared to that of Zephyros smiting and scattering clouds brought up by Notos (Il. xi., 305). In this last passage, and in Il. xxi., 334 and elsewhere, the word Argestes (which later on became a specific name of a distinct wind lying between Boreas and Zephyros), is used as an epithet of Notos; indicating (probably) that this wind is accompanied by bright cumuli and not by an overcast sky. This confusion of attributes may, to a great extent, be due to the want of further subdivision.

Hesiod (circa B.C. 735) names Notos, Boreas, and Zephyros only; and he speaks of them as beneficial winds; but of the rest (without names or number) as mischievous. The former are "the children of the morning," which may mean that they arise at the dawn; while the others, "random breezes," are the children of Earth and Tartaros (see Hes. Theog. 378-380:

869-880). He assigns the epithet Argestes to Zephyros; and not, as Homer did, to Notos.[1]

It can hardly be supposed that Hesiod included Euros among the "random breezes"; and it is difficult, if not impossible, to account for the omission of Euros from "the children of the morning." That Hesiod adopted a merely tripartite division of the heavens seems quite inadmissible.

Between the age of these earlier poets and that of the Philosophers, eight principal winds at least had acquired definite and specific names. That these names were in common use, and were adopted in, and not the creation of, the Schools is obvious from the archaic and almost rude forms of the names themselves; so archaic in fact that in many instances their derivation and meaning are merely matters of speculation. But it is in the Aristotelian philosophy that we first find an attempt to define accurately the direction of the winds on a scientific basis.

Theophrastus nowhere in the foregoing works defines the direction of the winds, but says (Weather signs, section 35), "the positions of the winds are such as are defined in the "diagram." This "diagram" to which he thus refers, is no doubt the same as that described in Arist. Meteor., lib. ii., cap. 6, where speaking of the winds, Aristotle says, "as to their position "we must consider the verbal description with reference to the "diagram . . . . Here the circle of the horizon is drawn.[2] Let "A be the place of sunset at the equinox; and, opposite to this, "B the place of sunrise at the equinox. Let another diameter "be drawn cutting AB at right angles, and G be the North, "and, directly opposite this, H be the South. Let F be the "place of sunrise, E the place of sunset, at the Summer sol-

---

[1] See note, infra., p. 94.
[2] See Figs. 1 and 2, and the explanations of those figures, infra.

" stice ; D the place of sunrise, and C the place of sunset, at
" the Winter solstice. Draw the diameters D E, C F.

" The winds are named according to their local position, as
" follows :—

" Zephyros—from A—that is sunset at the equinox.

" Apeliotes—from B—that is sunrise at the equinox and
" opposite to A.

" Boreas and Aparctias—from G—the North.

" Notos—from H—the South.

" Kaikias—from F—sunrise at Summer solstice.

" Lips—from C—sunset at Winter solstice.

" Euros—from D—sunrise at Winter solstice.

" Argestes (otherwise Olympias or Skiron)—from E—sun-
" set at Summer solstice.³

" These winds are opposed to each other in the direction of
" diameters of the circle ; but there are others which have no
" opposite winds, namely,

" Thraskias—from I—that is between Argestes and Aparctias.

" Meses—from K—that is between Kaikias and Aparctias.

" The line drawn from I to K is practically in the direction
" of the Arctic circle,⁴ but it is not exact. But there is no
" wind opposite to Meses—that is from M ; nor to Thraskias—
" that is from N, except that from N and over a small area a
" wind does blow, which the people there call Phœnikias."

According to this definition the direction of four of the winds
(that is Zephyros and Apeliotes and, by reference to the direction of these, Boreas and Notos), is determined by the places of

---

³ This wind is also called Iapyx in Arist. de Mund. iv., 12; and see Hor.
Od. i., 3, 1 ; Virg. Æn. viii., 710.

⁴ This line is called in Arist. Met. ii., 5, ὁ διὰ παντὸς φανερός, and in Met. ii.,
6, ὁ διὰ παντὸς φανόμενος ; each expression meaning " the line of constant
" visibility"; i.e., within which (for a certain part of the year) the sun never
sets.

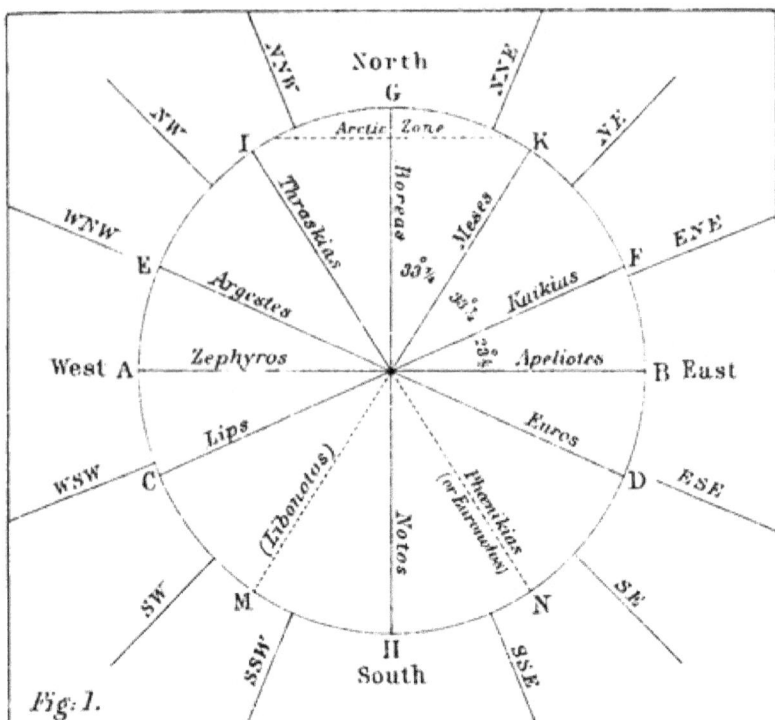

Fig. 1.

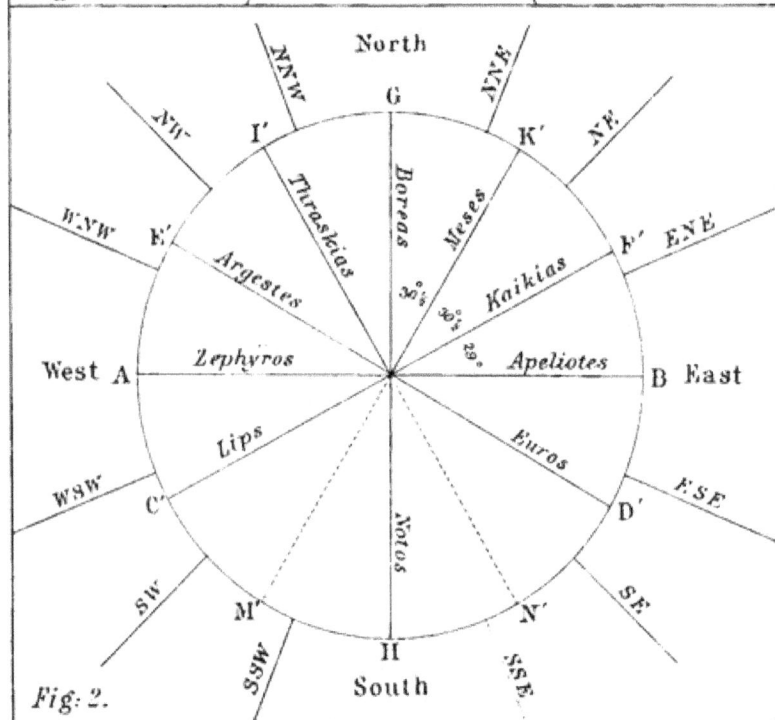

Fig. 2.

sunrise and sunset at the equinox. Those places are due E. and
W. of an observer on whatever parallel of latitude he may be,
as long as it be not too near the pole for the sun to set at all.
The direction of these winds therefore may be definitely referred
to the points W. E. N. S. respectively.

But of the next four principal winds, the directions of Kaikias
and Euros are determined by reference to the places of sunrise
at the summer and winter solstices; and those of Argestes and
Lips by reference to the places of sunset at those times.

Now the angles which lines, drawn from any place of observation to the places of sunrise and sunset, make with a line
drawn E. and W. through the place of observation, vary as
the distance of the latter place from the equator increases. To
a person situated on the equator the sun appears to rise, and set,
at the solstices 23° 27' North or South (as the case may be) of
the E. and W. points respectively. On the parallel of Athens
those angles are increased to 29°; and on the parallel of Greenwich to 40° 30'.

This will be understood more clearly by considering the
accompanying diagrams Fig. 1 and Fig. 2—in which the
lettering corrresponds to that in Aristotle's diagram above
described.

In each of these the line A B is drawn W. and E. through
the place of observation : G H is drawn N. and S. These lines
are constant. In Fig. 1 the observer is supposed to be on the
equator : in Fig. 2 he is supposed to be at Athens. Consequently, in Fig. 1, D (the place of sunrise at the winter solstice)
is almost E.S.E. In Fig. 2, D' has moved so as to be nearly
midway between E.S.E. and S.E. If a similar drawing were
made for Greenwich, D would be close to S.E.; and to an
observer still further North, D would approximate nearer and
nearer to S.

Similar observations apply, of course, to C, E, and F, in

Fig. 1, which in Fig. 2 assume the positions C', E', F',
respectively.

It is impossible to suppose that Aristotle intended to convey
that the directions of these four winds are variable, and that they
depend upon the situation of the observer. At the same time
he has not told us on what parallel the winds would have the
directions which he defines. For several reasons I think we
must not place his observer at the equator. He holds that the
S. wind does not come from south of the equator, but from the
northern half of the torrid zone, and that a corresponding wind
of the same character rises in the southern half of that zone
and goes southward. So that his diagram would not be true in
that respect if the line AB be on the equator. But he gives us
no other information about the equatorial region; and altogether
it seems most reasonable to suppose that, writing at Athens for
Greeks, he took Athens as his centre of observation.

In this view Fig. 2 (and not Fig. 1) accurately represents
Aristotle's diagram with the addition of the compass points.
The result is that Euros, Lips, Argestes, and Kaikias do not so
nearly approximate to the points E.S.E., W.S.W., W.N.W.,
and E.N.E., as they would do if we could accept Fig. 1 as the
basis of Aristotle's definition; but for reasons of convenience I
have taken those points as representing with sufficient accuracy
the direction of those winds throughout the foregoing
translations.

The directions of the other winds, Thraskias, Meses, &c.,
being determined by reference to the four just mentioned, the
same observations apply in their case.

At this stage, therefore, while Boreas, Zephyros, and Notos
maintained their original and cardinal positions, Euros had
become supplanted by Apeliotes, and had itself been assigned a
position southward of E.; the division of the horizon was not
into equal parts; and of the winds named by Aristotle, eight

occur in opposing pairs; two (Thraskias and Meses) appear to have been admitted almost on sufferance to further sub-divide the two northern quarters; while in the southern quarters. Phoenikias was regarded as a merely local wind not worthy of a place, and Libonotus was as yet unrecognized.

Aristotle treats Boreas and Aparctias as synonymous, and places them due N. But there are indications in Theophrastus that he considered them to differ slightly, although still keeping Boreas, as opposed to Notos, in the line of the meridian.

I have therefore (in my translation) placed Aparctias as a separate wind at " N. by W.," a position relatively, though not absolutely, maintained in later times; for, when Aparctias afterwards ejected Boreas from N., the latter shifted to the eastward.

The shifting in later times of both Boreas and Euros in the same direction from their original cardinal positions is not unremarkable.

Reference should here be made to the table in the Introduction (p. 17), showing the points to which each wind named by Theophrastus is referred in the foregoing translation; which either actually or approximately agree with the points ascertained by reference to Fig. 2.

We now pass on to perhaps the most interesting scientific monument of the ancient world. On the north side, and nearly at the base, of the hill which was crowned by the Akropolis of Athens, stands the octagonal building sometimes called the " Tower of the Winds," but more properly the " Horologium " (or Time Indicator) of Andronikos Kyrrhestes, by whom it was erected in the second century B.C. A full description of this splendid monument, with accurate scaled drawings of all its parts, will be found in the 1st Vol. of Stuart and Revetts' Antiquities of Athens (London, 1762). From this work, and from Vitruvius who described it (lib. i., cap. 6), it appears to

have served several purposes. Primarily it was to measure time,
and mark the diurnal and periodic movements of the sun by
dials, incised on its eight external marble walls, and still to be
seen. It stood in the line, or at the end, of a conduit from a
spring on the slopes of the Akropolis, the water of which filled
the basins of a clepsydra or water clock, in the basement of
the Tower, to register the hours between sunset and sunrise,
or on cloudy days. It also probably served the ordinary purposes
of a fountain to supply water to the inhabitants of that part of
the city.

But the important features for our present purpose are that on
the frieze are carved eight winged human figures, each sym-
bolical of a wind and its attributes; and above each just under
the cornice is inscribed the name of the wind so represented.

" Under each of these figures there is a sundial; and as the
" east dial is only the west reversed, and as the noonday line
" on the south dial is a perpendicular, from which the hour lines
" belonging to the forenoon are equally distant with the corres-
" pondent hour lines belonging to the afternoon, it is obvious
" that the astronomer who marked out these dials supposed the
" sides of this octogon (sic) tower exactly fronted the four car-
" dinal points of the horizon, and the four principal intermediate
" points. And it appears he was not mistaken; for on apply-
" ing to its western side (which according to this supposition
" should be in the plane of the meridian) a magnetic needle
" · · · · · it deflected from this side to the west about 12° 55':
" which, as far as could be ascertained by repeated meridian
" observations of the sun, was at that time the magnetic
" variation at Athens." *

* Stuart's Athens, Vol. I., p. 14. See also a paper on this tower by Dr. G.
Hellmann in " Himmel und Erde," II Jahrgang, 3 & 4 Heft., 1890. From
photographs in my possession it seems that many of the names on the
cornice have almost, if not quite, " weathered out " since Stuart made his
drawings.

The importance of this fact is that, as Notos is represented and named on the frieze above the face on which "the south dial" is described, and Boreas on the face parallel to it, we find that Boreas at the date of the erection of the tower still maintained its original direction from due north; and the line connecting Boreas and Notos is in the true line of the meridian.

The general design of the building will be seen from the photographic representation facing p. 77; which, owing to recent excavations, shows a great deal which Stuart had only conjectured, although with perfect accuracy. It is sufficient to say that the tower is a little more than forty feet high to the top of the cornice; and twenty-five feet through from face to face.

Vitruvius (lib. i., cap. 6) tells us that originally a brazen Triton stood on the apex of the roof, capable of rotating with the wind; the direction of which it indicated by pointing with a wand in its hand to one of the figures on the frieze below. This, we learn from Stuart, had disappeared; and in the cavity where it stood the Sheik Mustapha had by way of ornament placed a large wooden model of his turban. This (or a reproduction of it) is visible in the photograph.

The winds are named and represented as follows, beginning at the North and passing round through the East :—

(1) Boreas—An old man very warmly clothed, holding a conch-shell.

(2) Kaikias—An old man with severe countenance, holding a shield with hailstones in it.

(3) Apeliotes—A young man with flowing drapery, holding in the folds of his mantle fruits, ears of corn and an honeycomb.

(4) Euros—An old man with morose countenance, much wrapt up, his mantle concealing his right arm and hand, and held up by the left to protect his face.

(5) Notos—A young man emptying a jar of water.

(6) Lips—A man of middle age bearing an "aplustre," that is
the ornamental finial of the stern of a Greek ship under
which the helmsman stood; thus indicating a fair wind
for navigation.

(7) Zephyros—A fair, almost effeminate, youth, nude except for
a loose mantle, the flowing folds of which are filled with
flowers.

(8) Skiron—This equivalent of Argestes, (see last note to
Theophrastus on Winds), is represented by almost a
replica of Boreas, except that he holds a large inverted
jar, very different from the water jar of Notos, which, as
Stuart suggests, may be a brazen fire pot indicating the
scorching quality of the wind and the lightnings which
attend it.

Of these figures, Lips and Zephyros alone have the feet bare.
Apeliotes has buskins without soles. All the rest have buskins
with thick soles.

The photograph shows the N., N.W., and W. faces of the
octagon; the figures of Skiron, and Zephyros being seen on the
frieze.

Whether the designer of this tower intended to indicate that
the winds do not come from definite points on the horizon (as
Theophrastus, following Aristotle, taught), but that each of the
eight winds has for its domain the arc of the circle subtended by
a side of the octagon, is perhaps uncertain; but I am much
inclined to think that he did, and that this tower marks an
epoch of change in the treatment of our subject.

It will be observed that, in any case, the four cardinal winds
retain their former position; while Kaikias, Euros, Lips, and
Argestes (Skiron) have swung further away from the equatorial
line, and taken up positions equidistant from the cardinal points.
Probably the difficulties attending the Aristotelian definition

had been felt in practice, and it had been thought better to
assume an arbitrary definition more symmetrical and certain,
rather than a merely theoretical one.

At the same time the intermediate winds, Meses, Phœnikias
(Euronotos), Libonotos, and Thraskias had disappeared.

It is now time to turn to Latin Authors to learn the terms
and positions assigned to the winds by them.

Passing by Varro and Vitruvius, each of whom has something to tell us on the subject, we find in Seneca (B.C. 5—
A.D. 65) a complete list of twelve winds as follows[6] :—

Septentrio.

Aquilo.

Caecias ...   . from Sunrise at Summer solstice.

Subsolanus   ...   ... „   „   „ Equinox.

Eurus or Vulturnus ... „   „   „ Winter solstice.

Euronotus.

Auster or Notus.

Libonotus.

Africus ...   . from Sunset at Winter solstice.

Favonius or Zephyrus ... „   „   „ Equinox.

Corus or Argestes   ... „   „   „ Summer solstice.

Thrakias.

We have here a list founded mainly on the Aristotelian
system, with Septentrio representing Aparctias: and Aquilo
representing Meses, or its later equivalent Boreas; and, if
the division is symmetrical, it follows, having regard to the
position of Favonius and Subsolanus, that Septentrio and
Auster are in the line of the meridian.

But Seneca considers that the line joining Thraskias and
Euronotus, and not that joining Septentrio and Auster, is in the

---

[6] Quæst. Natur., lib. v., cap. 16.

line of the Meridian Axis.[7] The effect of this is to thrust off Boreas (in the person of its equivalent, Aquilo) yet another place from the meridian line; or perhaps we should rather say to remove the meridian line further from Boreas. The change is a singular one in many respects, and involves considerations which cannot be gone into here.

It is, however, remarkable that Seneca's meridian would approximate to the magnetic.

Pliny the elder (A.D. 23-79), in his Natural History (lib. ii., cap. 46) gives the following list of eight winds :—

Septentrio ... N.

Aquilo ... between N. and sunrise at Summer solstice.

Subsolanus ... from sunrise at Equinox.

Vulturnus ... ,, ,, ,, Winter solstice.

Auster ... S.

Africus ... from sunset at Winter solstice.

Favonius ... ,, ,, ,, Equinox.

Corus ... ,, ,, ,, Summer solstice.

The division is not symmetrical, and he has no wind from sunrise at the Summer solstice. He mentions, however, that some add to the list Thraskias, Cæcias, Phœnikias, and Libonotus; and also Euronotus, not identifying the latter with Phœnikias. But there is no indication of his adopting Seneca's view as to the meridian, or of Auster being otherwise than in that line.

On the Belvidere Terrace, adjoining the Museo Pio Clemen-

---

[7] " A Septentrionali latere summus est Aquilo; medius Septentrio; imus Thrakias. . . . . A meridiano axe Euronotus est; deinde Notus (Latine Auster); deinde Libonotus." The latter sentence is clear in its terms. " Euronotus " is from the meridian axis." The former sentence is not so plain; but " imus " must refer to the lowest (apparent) position of the sun, as it passes round by the North, which is of course in the meridian line; and there Seneca places Thrakias, with Septentrio and Aquilo next in order as the sun rises thence on its path towards the East.

TABLE OF THE WINDS IN THE MUSEO PIO CLEMENTINO.

tino of the Vatican, stands what has been termed the "Table of the Winds." It is in fact a flat-topped block of stone resting on a circular base. The upper part measures about 24 inches across and is about 18 inches high. It has twelve vertical faces or sides separated by projecting flutings at each angle. There is a slight depression on the top about the centre in which possibly a wind-vane may have stood.

On the upper surface, close to the edge above the face numbered 1 in the following list, is inscribed "Septentrio" (North); similarly above face 4 is inscribed "Oriens" (East); above face 7 "Meridies" (South); and above face 10 "Occidens" (West).

On the twelve faces are cut the names of twelve winds in Greek and Latin as follows:—

1.	Aparkias	Septentrio.
2.	Boreas	Aquilo.
3.	Kaikias	Vulturnus.[8]
4.	Apheliotes...	Solanus.
5.	Euros...	Eurus.
6.	Euronotos	Euroauster.
7.	Notos...	Auster.
8.	Libonotos	Austroafricus.
9.	Lips	Africus.
10.	Zephyros	Favonius.
11.	Iapyx...	Chorus.[9]
12.	Thrakias	Circius.

We have here a list differing from Seneca's list, so far as regards names, only (except for the mistake to be mentioned presently) by the introduction of the new terms: Euroauster,

---

[8] As to this error in the "Table," see later on.

[9] This is a mistake of the sculptor for "Corus." In several classical writers the name is spelt "Caurus." "Apheliotes" and "Thrakias" are, of course, only dialectic variations.

Austroafricus and Circius. The positions differ from those assigned in the Aristotelian system and by Seneca, in that each wind has assigned to it an arbitrary and equal division of the circle. The Greek names differ from the Aristotelian in that Aparetias (written Aparkias) has finally parted company with, and assumed the place of, Boreas ; and the latter is practically relegated to the place of Meses, which disappears.

A curious error is observable in this "Table," in that Vulturnus is identified with Kaikias instead of with Euros. That Vulturnus blew from the South of East is clear from the passages of Seneca and Pliny above referred to ; and from Aulus Gellius ;[10] and the name is derived from the Mons Vultur in Apulia which lies to the S.E. of Rome. Clearly, therefore, if the sculptor of the table had used the name Vulturnus at all, it should have been as a synonym for Euros, and he should have given the Latin form Cæcias (as used by Seneca and Pliny) as the synonym for Kaikias. We shall see that the mistake was perpetuated by later writers, a fact which seems to indicate that the Roman "Table" was well known.[11]

This "Table" was found in 1779 at the foot of the Esquiline towards the Colosseum in the garden of the Monks of Mount Lebanon.

Signor Isidoro Carini of the Vatican Library has been good enough to examine the characters of the inscriptions, and

---

[10] Noctes Atticæ, lib. ii., cap. 22, where however Gellius distinguishes Eurus from Vulturnus, restoring the former to E. He wrote about A.D. 143.

[11] That "Vulturnus" was an old-established name is shewn by the remark of Seneca that its Greek equivalent Eurus " has now become naturalized with us ; " " Eurus jam civitate donatus est."   But Livy, writing about the same period, when telling of the disastrous effect of the dust blown by Vulturnus into the eyes of the Romans at the battle of Cannæ (lib. xxii., §§ 43, 46), speaks of the name as a local term used in Apulia. The derivation recently proposed by Dr. Undauft, " from 'Vellere-vulsi,' as indicative of a tearing rapacious wind," is inadmissible as inconsistent with the character of the wind, as well as on other grounds.

expresses his opinion that it is certainly not older than the 2nd or even the 3rd century of our era. It thus marks a later development of wind nomenclature than the Athenian Tower.

Through the courtesy and kind assistance of Padre Francesco Denza, the Director of the Specola Vaticana, photographs of the "Table" have been obtained, a copy of one of which will be found opposite page 89.

At this stage it will be convenient to present in a tabular form a comparison of the positions or range of the Winds according to the Philosophers, the Tower of Andronikos, and the Vatican Table respectively. The degrees are measured in the direction from the N. through E.

Greek Name.	Latin Name.	Aristotle.	Tower of Andronikos.	Vatican Table.
Aparctias	Septentrio ...)	0°	-22° 30' to 22° 30'	(-15° to 15° (15° to 45°
Boreas	Aquilo ......)			
Meses	....	33° 15' ±		
Kaïkias	Cæcias (or incorrectly Vulturnus)..	66° 30' ±	22° 30' to 67° 30'	45° to 75°
Apeliotes	Solanus (or Subsolanus..	90°	67° 30' to 112° 30'	75° to 105°
Euros	Eurus or Vulturnus	113° 15' ∓	112° 30' to 157° 30'	105° to 135°
Phœnikias (or Euronotos) ..	Euroauster ...	146° 30' ∓	....	135° to 165°
Notos	Auster	180°	157° 30' to 202° 30'	165° to 195°
Libonotos (of Theophrastus)	Austroafricus	213° 15' ∓	....	195° to 225°
Lips	Africus	246° 30' ∓	202° 30' to 247° 30'	225° to 255°
Zephyros	Favonius ...	270°	247° 30' to 292° 30'	255° to 285°
Argestes (Skiron, Olympias or Iapyx) ...	Corus	293° 30' ±	292° 30' to 337° 30'	285° to 315°
Thraskias	Circius	326° 45' ±	....	315° to 345°

The later Greek writers in the following centuries adhered mainly to the Aristotelian divisions and names. Agathemerus (circ. A.D. 250) gives a list of eight winds and their places, obviously taken from the above quoted passage of Aristotle, and repeats that "Notos and Aparctias are opposed." He mentions, however, that Timosthenes the Rhodian (circ. 282 B.C.), made up twelve winds by adding the other four mentioned in the last table; except that he put Boreas for Meses and omitted the latter.

Adamantius, a Greek Physician, wrote (about A.D. 415) a Greek treatise on Winds founded on, and in places little altered from, the Meteorologica and Problemata of Aristotle and the works of Theophrastus. He accepted the duodecimal division, and the names given by Aristotle.

S. Isidore, of Seville (A.D. 560-630), in his Etymologies (lib. xiii., cap. 11) describes twelve winds, giving the same Latin names as those on the Vatican Table; even repeating the mistake as to Vulturnus, treating it as distinct from Euros and lying to the N. of E. Arevali, the learned editor of this work (Rome, 1801), noticed this discrepancy between Pliny and S. Isidore, but says that he cannot account for it. It seems extremely probable that the Archbishop had seen the Table on the Esquiline and had copied from it.

The Emperor Charlemagne is generally credited, on the authority of Eginhard (Vita Karoli Imperatoris, cap. 29), with the introduction of a nomenclature which laid the foundation of that now in use. It may be doubted whether this may not rather be due to the learned Alcuin, who was born at York A.D. 735 and in A.D. 782 went to France at the Emperor's solicitation to promote scientific learning in that country, and died Abbot of S. Martin of Tours A.D. 804.[12]

---

[12] Guizot, "Histoire de Civilization," ii., 176.

However this may be, Eginhard (p. 92, ed. Teulet) tells us that the Emperor "gave distinguishing names to twelve winds: "while up to his time it was scarcely possible to find names "for four" (probably meaning that there were not Frankish names for more), "and re-named them as follows:—

" Subsolanus	... he called	Ostroni	wint.	(E.)
" Eurus	... „ „	Ostsundroni	„	(E.S.)
" Euroauster	... „ „	Sundostroni	„	(S.E.)
" Auster	... „ „	Sundroni	„	(S.)
" Austroafricus..	„ „	Sundwestroni	„	(S.W.)
" Africus	... „ „	Westsundroni	„	(W.S.)
" Zephyrus	... „ „	Westroni	„	(W.)
" Corus	... „ „	Westnordroni	„	(W.N.)
" Circius	... „ „	Nordwestroni	„	(N.W.)
" Septentrio	... „ „	Nordroni	„	(N.)
" Aquilo	... „ „	Nordostroni	„	(N.E.)
" Vulturnus	... „ „	Ostnordroni	„	(E.N.)"

In all these "old-high German" names the termination "roni" means "running from": and is the origin of our termination "ern" in "Northern," &c. See Prof. Skeat's Etym. Dict., s.v., where he says the derivation of "North" is unknown. "East," according to the same authority, indicates "the place of shining, or the dawn"; "South" (or Sunth), "the sunned quarter"; and "West" the "resting or lodging-"place."[13]

---

[13] Since this Appendix has been in type, I have had an opportunity of perusing a paper, "Ueber die namen der Winde," by Dr. Friedrich Umlauft, in the January, 1891, number of the "Meteorologische Zeitschrift," reprinted from the "Deutsche Rundschau für Geographie und Statistik, vol. xvi., No. 3. In it he has traced very slightly the changes of names and positions; the purpose of the paper being to explain the meaning and determine the derivation of the names used at various times and in various countries of as well the new as the old world. Such an enquiry belongs rather to the region of philology than meteorology; but I am compelled to say that some of his derivations, though plausible and ingenious, must be accepted with caution, and are

In this list the error of the Vatican Table is again repeated,
" Vulturnus" being assigned to "Ostnord," or East-North,
which would again indicate that Charlemagne (or Alcuin)
derived the list either directly from Rome, or indirectly through
S. Isidore.

A similar combination of the names of the cardinal points
to denote the intermediate points had come into use in this
country in the time of Archbishop Alfric (circ., A.D. 995), who
gives in his Vocabulary[14] the Anglo-Saxon equivalents (such as
Norðan-Eastan wind, Suðan-Eastan wind, &c.) for the Latin
names of the Vatican Table. Unfortunately, however, the
list as it appears in the Bodleian MS. is full of obvious copyist
blunders. Amongst others Euros and Euroauster are coupled
together (with the result that only eleven winds are accounted

rather speculative. After mentioning Homer's four winds, Dr. Umlauft says,
" Hesiod also recognised these four chief winds; but named Argestes in the
place of Euros." This is a mistake; but Dr. Umlauft is not alone in it. It
will be found that, as I have stated in the text (p. 79), the term Argestes is
used by Hesiod, as by Homer, as an epithet of one of the winds, and not as
the name of a wind. He in fact names three winds only, as above stated.
This is plainly seen if due attention is had to the position of the conjunctions
in the places where the names occur. The paper however, as well as a
review of it in the American Meteorological Journal, vol. xi., p. 67, will well
repay perusal and consideration.
I am indebted to my friend Dr. Isambard Owen and, through him, to
Professor Morris Jones, for the following note on the Welsh names of the
cardinal points:— " Deheu" (South) means "to the right hand," the spec-
tator being supposed to face the rising sun. Similarly "Gogledd" (North)
means " leftwards." " Dwyrain" (East) means " rising," from root "dwyre."
"Gorllewin" (West) is said to be doubtful, but is attributed to " gor,"
signifying " beyond:" and " llewin," from a root signifying " light." If
I might hazard a conjecture, I would suggest that "Gorllewin" is a con-
tracted form for " gorchllewin," which would mean the " enclosure" or
" folding place" of " light:" representing the same idea as " West" as
explained as above by Professor Skeat. I have it also from the same
authorities that the Welsh named the winds simply after the cardinal points,
except that the East wind was called " Gwynt traed y meirw," or " the
wind of the feet of the dead;" so called because the feet of the dead
when buried are turned to the E.; and is also called " Gwynt y rhew," or
" the frost wind."
[14] Anglo-Saxon Vocabularies by Thos. Wright, ed. Wülcker (1884). Vol. I.,
pp. 143-144.

for) and assigned to Norðan-Eastan. Vulturnus (probably by accident, as Cæcias is not named) appears in its right place as Eastan-Suðan. Mr. Wright, in his suggested correction of the MS., would have repeated anew the error.

Bartholomew, an English Friar, in his "Prohemium de proprietatibus rerum" (Lyons 1480), has a dissertation on the air and winds. He retains the duodecimal division; dividing the winds into four principal winds, each having two collaterals. To the E. wind he assigns Vulturnus towards the N. (the old mistake) and Eurus towards the S. as collaterals; to the W. wind (which he calls Favonius) he assigns as collaterals Circius (a mistake for Corus) towards the N., and Zephyros (a mistake for Africus) towards the S. To the S. wind he assigns Notos and Africus (mistakes for Euroauster and Austro-africus respectively); and to the N. wind (which he calls Boreas) he assigns Aquilo towards the W. and Corus towards the E. (mistakes for Circius and Aquilo respectively) as collaterals. A work so full of blunders is only worth noticing to prevent their being repeated.

Joachim Camerarius, otherwise Leibhard (b: at Bramberg, A.D. 1500; d: at Leipzic, A.D. 1574), wrote two metrical works in elegiacs; one Æolia, "On the names and places of "the winds;" the other Prognostica, or "Weather Signs" (Nuremberg 1535). The former recounts the various divisions into 4, 8 and 12, and traverses generally the ground we have passed over in dealing with them; including a brief reference to the Tower of Andronikos; but he has not added to our stock of information. At the end of Æolia however he has given three diagrams to illustrate the three divisions; but as in all the eastern half of the circle is on the left and the western on the right, they are very difficult to follow. He seems to have intended them to be diagrams of the heavens viewed from below. In that of the duodecimal division he puts Aparctias

at N. and Boreas and Meses (as identical) at the next point
eastward. In his diagram of the "8" division he has, clearly by
mistake, put Lips at N.W. and Argestes at S.W.

We may conclude our retrospect with a reference to Vin-
cenzo Coronelli (1693), who in his "Epitome Cosmografica"
gives several diagrams. The first is of the duodecimal division
with Greek and Latin names: in which the only important
variations from the Vatican Table are that he gives Aquilo as
the equivalent of Aparctias at N.; ignores the names of Boreas
and Septentrio; and restores Meses to its place in the Aristotelian list, and Vulturnus to its proper place as the equivalent
of Euros. He next gives a diagram of thirty-two winds with
Greek (and some Latin) names, many of which are evidently
of late composition. It is sufficient to give those in the first
octant—

N.	Septentrio or Boreas.
N. by E.	Hyperboreas or Gallicus.
N.N.E.	Boreas or Aquilo.
N.E. by N. ...	Mesoboreas.
N.E.	Arctapeliotes.

He also puts at S.W. Notozephyrus. It is obvious that no
reliance can be placed on such a list as authentic, and that an
attempt was being made to give apparently classical names for
the more modern division.

He also gives a diagram of sixteen winds with names in French
and Italian; and finally three diagrams in which the thirty-two
points of the mariner's compass are set out, with the names in
Dutch, English, and Italian respectively; the English terms
being those now in use, except that "North to East" appears
instead of "North by East."

When the division of the horizon into thirty-two parts was
first adopted will probably never be known. It was the natural
result of constantly dividing the original quaternal division by

two. We learn from Chaucer[15] that it was in use at the latter end of the fourteenth century for nautical purposes, while twenty-four divisions were made for astronomical purposes. It is worthy of remark that the Chinese compass in use since the fifth century has twenty-four divisions only.[16]

If my readers have followed me so far, they will come to the conclusion at which Aulus Gellius arrived centuries ago, when he wrote that there was no general agreement either as to the names, the position, or the numbers of the winds.[17]

J. G. W.

---

[15] " Now is this Orizonte departed in xxiiii partiez by this azymutz in significacion of xxiiii partiez of the world ; al be it so that ship men rikne thilke partiez in xxxii." *On the Astrolabe*, quoted in Encycl. Brit.: tit., Compass, Mariners'.

[16] Encycl. Brit.: ubi supra.

[17] Aul. Gell.: Noct. Att. lib. ii., Cap. 22, " Quia vulgo neque de appellationibus ventorum, neque de finibus, neque de numero conveniret."

KENNY & Co., PRINTERS, 25 CAMDEN ROAD, LONDON, N.W.